The Cure

A Novel

I0471272

By Durwood White

For Mary Bell

My thanks reach out to this avid reader of novels; one of sensitivity to details and viable plots.   She often redirects my thoughts when my mind sticks in some nebulous cloud of darkness and doubt.   Thank you, Mary.

Acknowledgement

Dr. AbulKalam Shamsuddin, M.D., Ph.D., and professor of Pathology at the University of the Maryland School of Medicine in Baltimore, has summarized his lifetime research in a book entitled, *IP6, Nature's Revolutionary Cancer-Fighter* (Kensington Publishing, 1998). He currently manages his company, IP-6 Research, Inc. I have borrowed heavily from his study in preparing the plot of this fictional narrative. Thank you, Dr. A. K. M. Shamsuddin for your kindly permission.

Chapter 1

MOUNT KILIMANJARO, the highest peak on the African continent at 19,340 feet, stood high in the glowing light of a September moon eternally crowned with snow. She sat proudly on the corners of west Congo and the southern corner of Zambia, quite a distance from the fossil sites at Olduvai Gorge, where Dr. Louis Leaky astounded the world with the discovery of ancient human bones. The sounds of the Indian Ocean tirelessly washed on the eastern coast of Tanzania, and the myriad of stars in the African sky were so brilliant they lit the Kenya border in a postcard presentation. An occasional meteor sliced across the heavens like a celestial knife, the Big Dipper so magnificent it obscured the huge planet of Jupiter. The evenings had its customary attack by bugs and gnats that fiercely dove into a victim's flesh; mosquitoes as big as a dragon fly. The distant drums of tribal ceremonies added an eerie atmosphere, as a lone laughing hyena sat on a distant hill silhouetted in the light of the full moon.

A twelve year old lad and his father, a prominent pathologist at a major medical center in New York City, left their camp in the misty dimness of dawn, and drove their Land Rover westerly across the Congo toward the impressive towering mountain. Although they had

4

calculated their arrival before sunset, somehow they had misjudged the persistent drag of the sandy terrain that sank the four-wheel drive to the axles because of the heavy equipment in the cargo bay. The two campers followed a map drawn on animal skin in India ink; old, perhaps sixty years, rumored that it had belonged to an obscure tribal leader in one of the villages in ancient Kenya.

The sun sat on the western horizon as they drove all day through the desert rifts and scrub trees. Numerous rocks gathered haphazardly on the piedmont floor, and the vehicle stalled, steam billowing from under the hood. An inspection revealed an overheated radiator, and the pathologist decided the remaining distance was within walking distance; he donned a backpack, and studied the map one last time.

Tall grass grew around the perimeter of Kilimanjaro as they walked through the saw-tooth maze parting the blades with their sunburned hands. The young lad followed behind, stretching his shorter legs as he stepped into oversized footprints. The heat was searing, and bugs were annoying but the lad trudged on like a trooper. This was the third summer trip to Kilimanjaro, and he suspected it would not be the last if they failed again; he loved these trips with his father.

After they had stalked through the tall grass for an approximate hour, the exhausted couple found green flora sprouting from black moist soil. The nutrient for this growth resulted from the life-giving snowmelt that flowed down the twin peaks of the mountain. Precipitation had settled layers of humus at on the piedmont floor, providing a soil enriched by decaying organic matter.

As the pathologist parted the shafts of grass, his eager eyes focused on the object he had dreamt for several years—a hut entangled in the embroidery of massive vines at the foot of Kilimanjaro. They had only two hours of daylight before sunset. Evening shadows would soon gather the splendor of twilight.

They must hurry.

Twisting among the snake-like vines, they forced their sweaty bodies through the rubble of empty wooden crates scattered over the overgrown premises. Nestled in a canopy of vines, the adventurers squeezed passed a broken wooden door precariously hung on one leather hinge. The shadows were eerie; cobwebs adorned the corners, and spiders occupied their webby traps for lunch. Giant cockroaches raced here-and-there around the one-room abode. The floor was a series of broken boards; the walls were once covered with reeds now raggedly split by the searing heat.

Dr. John Cook panned his flashlight around the once spotless laboratory now scattered with broken bottles, some marked with the skull-and-crossbones; not pirate stuff, but the symbol of concentrated acid. A hand-turn, two-position centrifuge was attached to a dusty table with a large thumbscrew. Petrie dishes were haphazardly scattered over a crude bench. A monocular microscope with exposed rack-and-pinion gears sat on a grimy table with glass slides randomly stacked in uneven piles. A cockroach crawled from inside the microscope tube wiggling its antennae, and scurried between the stacks of slides. A wire cage located next to the microscope had probably housed live specimens for research as evidenced by several skeletons of mice and a few chicken carcasses. A black Mamba snake grossly crawled through the spokes of a cage and slithered through a gaping hole in the wall.

Cook's roving eyes seized on a tattered leather notebook that lay by the microscope smothered in powdery filth. He carefully opened the aged book. Stiff pages fell open at a worm-eaten bookmark. A hand-sketched diagram scrolled in India ink flooded Cook's awareness.

The faded notes were smudged and pale, but Cook recognized the structural shape of an organic compound: some type of sugar derivative. Quickly he deciphered the six radicals attached to the double bond carbon-ring: hexaphosphate, he reasoned. His mind suddenly screamed: phytin—phytate! *That's it!*

Mental computations quickly deciphered the cytology of the chemical structure as he gazed at the notations. This primitive scientist had recorded fascinating data *in vitro* testing with mice and chickens. Before Cook's hypothesis gripped his mind, a cursory reading of the test data exploded his theory!

Somehow it all became clear.

This genius had isolated phytate, the salt of inositol, and had tested it on . . . *what's this*—humans?

The test results were electrifying.

The young twelve-year old lad could no longer contain his anxiety. "What is it, father—what's the meaning of that diagram?"

The professor heard the question but said nothing, his mind captivated by the handwritten notes as he turned page-after-page of experimental history.

Finally he gently closed the notebook, and gingerly slipped the priceless item into a satchel that he had brought for such an occasion. He slowly released his abdomen muscles in a controlled sigh, and answered his son's question.

"Dynamite, son—this notebook is dynamite."

The lad's eyes enlarged beneath his wrinkled brow, his young active mind suspended in rapid calculations, as he dutifully followed his father outside the dark, dank hut.

"But whose laboratory was this, father?" the lad insisted tugging at his father's baggy trousers.

Cook squatted and gripped the lad by both shoulders. "Chip, I didn't tell you, but the skeleton of the scientist is lying on the floor, its head removed."

His face grimaced. "Who ran this lab?" the lad persisted.

"I don't know, son. He didn't record his name on any of the pages that I saw. But this lab was in operation at least by 1929."

"How do you know that, dad?" he wondered.

He smiled at his inquisitiveness. "I saw a wooden crate with an address painted on the slats, son."

"Address?"

"Yes, I think this researcher may have ordered inbred strains of mice from Jackson Laboratory in Bar Harbor, Maine, a company founded in 1929. It's now the largest supplier of research mice."

"Then we can trace the man," the lad reasoned.

"Good observation, Chip," he said smiling as he kneeled and laid his big hand on his son's sloping shoulder.

"Son, I am very proud of your interest in biology and physics. Someday you will be a good physician."

Chip heard his father's words, and yet he wasn't sure his mother agreed with a physician's profession. There were many things he wanted to do with his life, but mostly he wanted his father to be at home more often.

Chapter 2

DR. JOHN COOK sat in his laboratory-office in the basement of the Cook Medical Center reading a tattered leather-bound notebook for the umpteenth time, while he consumed a pot of coffee. He had studiously decided the tests described on the wrinkled pages were performed between 1921 and 1930, according to the last entry noted as August 6, 1930— plus the fact that laboratory mice were not available before 1921. The identity of the genius who wrote these notes and conducted these tests had enamored his thoughts over the past weekend. He pushed his spectacles to the top of his head and mused of the fruitful trip to Africa in the summer of '79. He had evaded his research for the last month owing to enrolling his son in the John Hopkins medical school.

Musing ceased.

The tattered pages of the laboratory notebook refocused in his eyes. Who was this prodigy of medicine, Cook wondered?

All day he had placed unanswered calls to Jackson Laboratory, and had finally reached a manager in the records department, although his answers were actually excuses. It became obvious that the old invoice records were sketchy, although the manager had expressed a word of encouragement

9

explaining that the formative years of records were archived in a warehouse. The company was founded in 1920, following the aftermath of World War I. The era was before computer filing, and it took excruciating hours of search before they located the handwritten invoice in question.

***

Cook sat at his IBM computer with its specially made 15-inch monitor, transferring data onto a cassette tape from his computer files. He had recorded dates, times, dosages, and results from weeks of concentrated tests, still without information from Jackson Laboratory.

Suddenly, the laborious task, compounded by long hours, averted his concentration, and he rocked back into his leather chair, snatched off his reading glasses, and dropped his hands into his lap. As he rubbed his tired eyes, observations stirred in his overtaxed mind. He fingered his chin as calculations rapidly streaked across his eye's retina. The mental conclusions erupted into a sudden chuckle. Uncontrolled laughter followed him to a microscope on the counter, and he eagerly peered through the dual eyepiece. The image on the slide became swiftly clear.

"Of course!" he shouted, as he joyously rocked back in his seat, folded his hands in his lap, tired eyes excitedly focused on the compound. The researcher had astoundingly isolated and transplanted human oncogene, a gene that caused cancer, into the embryo of an inbred mouse! He had recorded therapy data on cultured *in vitro* cancer cells by injecting the embryo with unpurified phytate.

The data was exhilarating, unbelievable, and too wonderful for realism in 1930. Despite unsafe conditions, using inferior equipment, this genius had stumbled upon a miracle cure for cancer, but could not reach an endpoint, shackled by the stigma of his insufficient technology.

Cook slumped into his desk chair as he slowly absorbed the gravity of this man's work. He was ahead of his time, unhampered by 20th century regulations, trapped in the dawning period of the industrial revolution, and strangled by a science more complex than his equipment's design; but not his tenacity, nor his dedication.

The telephone buzzed and scrambled is thoughts.

Cook's eyes squinted as his brain traveled from a tiny lab in East Africa in 1930 to a modern laboratory in New York City in 1990. He punched a button on the console.

"This is Dr. Cook," he wheezed.

"I've found something, Dr. Cook," replied an anxious voice that belonged to the manager of Jackson Laboratory."

"That's wonderful!" Cook exclaimed, laying his reading glasses atop the data pages.

"Your man was a Doctor Michael Shannon of New Zealand. He ordered three-dozen inbred mice—one of our first strains in March of 1929. It was shipped to Cape Town, Africa."

Cook's eyes squinted shut. "Makes sense—good work! You may bill me for your time . . . may I impose upon you again, and ask if you would send me a copy of that invoice?" he asked with relaxed excitement, his mind still chewing on the accomplishment of Dr. Shannon.

"Can do, right away, Dr. Cook."

Cook punched a button and ended the conversation, his mind still seized by the message: *"can do."* If only a cure was possible in his lifetime, he thought from his 20th century training.

His racing mind finally completed the calculations, a dozen research angles spread on the screen of his retina. Then suddenly his thoughts exploded in a myriad of lights.

"Eureka," he barked!

He suddenly recalled that Shannon had used natives from the nearby villages as human test specimens–human data! His deduction had probably accounted for the researcher's violent death. Perhaps a disgruntled witchdoctor had influenced the ignorance of the tribal men.

"Who knows," he thought aloud. "What if he operated a hospital and had found a cure for cancer in 1930!" His voice echoed in the empty room.

Chapter 3

All week Dr. John Cook had felt that someone had watched his every move, and yet he thought it only an image in the shadows of his tired mind, too, it had reminded him of a recurring rumor that pharmaceutical companies had planted spies in every research institute in the country. The notebook of Dr. Shannon and his research were priceless, and he was satisfied that he had arranged adequate protection by encryption.

After Cook had returned from Africa, he drew up a will that left all his research notes and recorded tapes to his son, Dr. Chip Cook, now in residence at Hopkins Hospital under Dr. Rosie Rothschild. He had every confidence that Chip would become a physician and could continue his work, and he knew his wife would grant his wishes. It was simply proficient planning that provided for the unknown.

He deeply sighed, rose from his chair, and walked the twenty steps to a bank-style vault that he'd installed for security. Cook strolled inside the steel unit as he had done on countless occasions, and placed the briefcase in a locked tray. Satisfied that the computer files were encrypted, the research was now secured in

13

the vault. He finally stepped out of the steel encasement and muscled the heavy door closed, spinning the locking wheel. A light on the wall blinked red," indicating that the door was locked.

Dr. Cook mused for a moment of how he had built Cook Medical Center from the roots of a small clinic as an ambitious young physician, the endless days of late hours and lonely nights until he had met his future wife. Jenny Lynn Bateman had approached him for a nursing position, and she was more than adequately qualified as an emergency room nurse. He hired her on the spot. They built the clinic together.

And then his thoughts drifted to a friend who had gone with him into Kenya, where they had met an old prospector who provided a map—the same map that he and his son followed when they discovered the old cabin at the foot of Mount Kilimanjaro. This friend had helped Cook finance the construction of Cook Medical Center, but somehow their friendship had grown aggressively cold over the years, an event sealed in the cortex of his mind.

His thoughts vanished when he realized he ran late for a dinner date with his wife and son, and he rushed to the lab door. As he flipped off the lights, stepped into the hall, turned, and rechecked the door lock with a twist of the inside knob, he felt a sudden crash on his head. Cook's limp body slumped on the floor unconscious.

Hands donned in black gloves gripped Cook's torso at the shoulders and dragged the body back into the lab area. The assailant removed his gloves and retrieved a case from his inside coat pocket. He opened the case revealing a syringe loaded with liquefied sarin. He had selected this covert poison because it was five-hundred times more toxic than cyanide. And it was undetectable by post mortem autopsy, since the high pH of stomach acid

14

decomposed rapidly into a non-toxic phosphoric acid, according to his meticulous research.

He quickly slipped on his gloves, rolled up a sleeve on the arm of the unconscious physician, and carefully removed the syringe from its case. The intruder gingerly snapped the safety from the needle tip. Gloved hands injected the poison into a vein on Dr. Cook's wrist. After he placed the syringe back into its securing case, he unrolled the sleeve, re-buttoned it, and leaned the body against the specimen cage. The assailant replaced the case in his inside pocket, assured that he'd left no evidence.

The man dressed in black moved to the door and switched off the lights. He disappeared down a narrow hallway on a preplanned route to a set of stairs leading out of the basement of Cook Medical Research Center.

The prominent physician's body began to twist and jerk as vomit poured out of his mouth, finally his blue lips drooled. Even comatose, the body had suffered a series of convulsive spasms.

John Cook lay dead.

Chapter 4

A LATE AFTERNOON overcast had precipitated a slow chilling drizzle that soaked the cemetery behind a Catholic church in downtown New York City. Cold rain relentlessly pelted the attendees. A crowd stood around a graveside randomly opening their umbrellas as the drizzle increased to a steady downpour. A minister read the last rights to Dr. John Cook. His son, newly entered into medical school, and the deceased's wife both sat in the first row of the graveside chairs under a protective burial canopy.

The muted hum of a DC motor broke the eerie silence as a dark-tinted window lowered on a black Cadillac Escalade parked by the curb some thirty feet from the graveside. A passenger sat in the inclement darkness of the rear seat wearing a raincoat and hat, face hidden in the shadow of its wide bream as cigar smoke hovered in high humidity. His eyes, covered by dark sunglasses, watched the men who escorted the bereaved wife and son to a waiting limousine. As the doors closed, the Escalade across the way beamed its headlights and motored away from the curb.

The passenger in the backseat puffed his cigar and blew a series of smoke rings that disintegrated on the neck of the driver. The countenance of the driver's face soured.

The backseat passenger growled, the cigar gnashed between his stained teeth. "You have got to get in that safe, understand—I want that formula."

The driver's fingers tightened on the steering wheel as he cursed under his breath. An excruciating pain suddenly exploded inside his head. He popped a pill into his mouth as they drove away.

***

On the backside of the Bronx, a Cadillac Escalade pulled to the curb and parked. The driver had driven the passenger to his home; good riddance, he imagined, yet he wasn't quite sure what he thought. Time, he needed time to gather his thoughts without ridicule and intimidation. His mind was disoriented by self-prescribed drugs, pills, and umpteen side effects that occasionally drove him mad. The driver's face suddenly wrinkled with piercing pain as pressure accumulated behind his forehead. He anxiously looked for the pill bottle he normally carried in his pocket, and fortunately spotted it on the floor. He stretched his arm to the floorboard, gripped the bottle, and hurriedly swallowed two pills, his heart racing, as perspiration poured down his tormented face.

Finally the pain gratefully subsided, his irritated eyes refocused, and he punched the keys on his cellphone. A circuit clicked, and then a raspy voice answered.

"Yes."

He drily swallowed. "We must get the formula," he said, his head spinning with pain.

A moment of silence persisted.

"Go to Plan B. And I suggest you engage your cousin."

He closed the cellphone, a thumb-and-finger pressed across his forehead like a vice. The weary man sat silently for a restless moment in deep thought, as his tortured mind busily calculated electron signals and sent them to degenerating synapses. Random signals somehow connected with his nervous system, impaired by alcohol and drugs.

Amazingly his mind cleared, if but a moment; the pain was apparently being controlled by the new medicine, he thought. He focused on Plan B: lay low, the son of Dr. John Cook was now the viable target. He sighed deeply and flipped open the cellphone. When the number finally focused in his mind, he punched it into the keypad, wondering if his cousin were home. The cousin had always been mechanically inclined, and was military-trained as a bomb expert, a feature of Plan B.

Chapter 5

DEEP IN THE basement of the Cook Medical Center a physician of thirty years peered through an electron microscope as he positioned a slide under the eyepiece. A picture of the slide flickered on a 21-inch monitor screen in vivid color. Entrenched in his mind were thoughts of his father's death and the encrypted cassette he had left in his will. After he had completed his internship, his mother had given him the cassette, which he copied onto a modern CD, and kept both files in the bank-style vault. The sample he was investigating was magnified by a Nikon lens; it was one of the many samples he had prepared from the information in the research notes left by his father.

A mature woman entered the lab dressed in a nurse's white uniform with a blue smock, a stethoscope draped around her neck signaled she was also a practitioner. A badge pinned to the lapel of the smock announced her name: Jenny Lynn Cook, Training Director. The young physician who sat staring at the monitor screen was her son, Chip Cook, whom she had never told him that John Cook was not his biological father. Indeed, John had died mysteriously although the coroner's autopsy was inconclusive, and he had ruled that a carcinogenic agent had infected him in his laboratory. Her feminine intuition convinced her that John's death was a cover-up; she firmly believed that someone on the Hospital Board of Directors was responsible for the murder of her husband. And she was so distraught by the circumstances that her caretaker advised her not to view his body prior to the funeral. Fortunately she had engaged her lawyers to prevent cremation, for which she had firmly disagreed because it would have destroyed any evidence.

But that was not her major concern. Her motherly instinct told her that Chip may discover that he was the son of a respected federal agent of the NSA in Washington attached directly to President Winston Darcy's staff. The federal agent was a former boyfriend Jenny Lynn had known long ago, longer than she cared to remember, but memories were difficult and painful to depress. He had also owned a laboratory in Jacksonville, Florida, and he, too, was clubbed on the head and left with his lab aflame. She had nursed the boyfriend back to health—the similarities were uncanny she suddenly realized. Was it possible that she had loved two men?

When her boyfriend was subsequently called to Washington by the President, she had lost him to another woman, a woman who was a close friend; a popular TV anchor in Washington. Jenny Lynn had left her hospital in Jacksonville, Florida, and moved to New York where she had worked in a clinic owned by Dr. John Cook. They were later married, even though she told him that she was pregnant. He had agreed to raise the child as his own son. Her distant memories, which she had relentlessly harbored, suddenly vanished as her mind returned to the present.

She deeply sighed, moved toward the desk, and stood behind her son, who watched the intriguing shadow of a molecule on the monitor of the electron microscope. She placed her slender hands on his broad shoulders.

"How's the research coming, Chip?" she whispered.

The handsome six-four M.D. swiveled in his chair and rubbed his fingers in his fatigued eyes, and then took his mother by the hand with a forced smile from lack of sleep.

"I've isolated the molecule—but that's all, I'm afraid. Dad's work gave me that insight—I've done nothing on my own," he admitted.

She sat down beside him and stared into her son's penetrating eyes until a recurring memory triggered a hidden image in the deep caches of her mind: her son had the same hazel eyes of her former boyfriend. Uneasiness forced her back to reality.

"Your father was a brilliant medical researcher— I'm so glad you have chosen to continue his work—it was his desire."

Chip inhaled and slumped heavily into the soft cushions of his leather chair. A flash from his active mind revealed the image of his father when they had gone on three summer safaris into Africa seeking a leather notebook. At his father's insistence, he had memorized the encryption code when a senior in High School. Suddenly his shoulders cringed as his head raised, the image replaced by urgent resolve.

"Mother, dad's work will continue—count on it! We will bring his killer to justice, I promise you that much," he replied resolutely.

Jenny Lynn gazed into space, pendant tears at the corners of her blue eyes, as she pondered his words in her heart. So, even Chip understood that his foster father had been murdered, although she had never told him he was adopted. And then a more terrifying thought entered her mind. If he had discovered the details of John's death, could he also uncover the fact that John was not his biological father? Her mind flashed back to the year Chip was born.

The year she left her boyfriend, the year she came to Cook Medical Center in New York City, the same year his son was born, and she'd buried her useless life in her work. She often frequented a café where she sat for hours in deep dismay, considering her biological condition with child. And yet a casual friendship developed when Dorothy Millhouse, a chorus-line dancer, who stumbled into the café one late weekend. She practiced dancing at a theater in the same block of the café, that is, when she had work. Jenny Lynn had seen her twice before, and during one eventful night the two lonely women had shared their deepest secrets around glasses of wine, a friendship sealed for all time.

Dorothy hailed from Biloxi, Mississippi, and ran away from home in search of a dancing career. Jenny Lynn was the ER head nurse at University Hospital in Jacksonville, Florida, and also ran away when she realized she was pregnant, and would never see her boyfriend again; surely, he'd forgotten her—but she'd not forgotten him. It was a puzzle that haunted her until the son was born. Just being around him gave her some peace, restful peace that lasted until she met with Dorothy again, and they talked through her problems. Dorothy was a God-send, and she needed a friend.

Chapter 6

CHIP HAD SPENT two exhaustive weekends locked in his basement laboratory deep in the Cook Medical Research Center Annex. Beverly Banner, a receptionist at the Research Center, had brought in food and drink that sustained his late hours in research. He had worked tirelessly night and day, dozing on his office sofa until he could work no longer. He rushed to complete a battery of tests using *in viva* cultures of K60 human leukemia cancer cells introduced in F344 mouse embryos, which were treated with IP6, followed by a duplicate battery of inositol plus IP6 two weeks prior to initiation with K60.

His eyes were glued to a digital stereo zoom microscope connected directly to the main computer including a built-in 2.0-mega piree camera with Motic imaging software. Chip studied the polarized image of a Myo-Inositol complex with intense concentration as he punched keys on the keyboard. He had programmed specialized medical research software specifically for this research project, and was excited at the prospect that the algorisms had worked perfectly. The software had searched its comprehensive library and had named the complex *Myo-Inositol Hexakisphosphate*. The chemical name was derived from its chemical structure: *inositol hexaphosphate* (IP6), soluble in water; thus a complex sugar.

"Holy Cow," Chip Cook whispered aloud. "This is one sweet compound."

His excitement derived from the chemical structure with its similarity to glucose. His keen mind suspected the injection of this sugar derivative actually reduced side-effects. He had recorded his observations in the databank of how the molecule experienced subtle changes in chemical structure. Each molecule added a phosphate (P) radical producing a family of inositol phosphates, each with its own ability to create biochemical effects in the body, the area of his final study—a coded *cure,* for lack of a more definitive word.

The computer software had named each phosphate derivative with code names according to the number of phosphate groups (P04): inositol mono-phosphate (IP1), inositol biphosphate (IP2), inositol tri-phosphate (IP3), and so on, through the six receptor sites of the inositol group.

This research was groundbreaking, the results monumental, and the investment capital astronomical. And although Chip knew he had significant statistical data, reproducible and consistent, a reduction in cancer cells that affected various tissues including the colon, breast, prostate, and liver—there seemed to be no end to its benefits, yet his limited work would not be acceptable by the FDA or the medical field. Why?

When a clinical study was statistically significant, it yielded test results that were accepted as valid by all researchers. If a study exhibited an effective treatment, yet had insufficient number of patients (sample size), it was insufficient for consideration as "statistically significant."

Chip raised his reading glasses to the top of his head rubbing his tired eyes. His mind replayed his conclusion: A significant number of patients—a viable clinical trial program, yet a potentially risky financial venture.

He gripped the mouse and clicked on "Print." A document rolled out of a laser printer containing twelve pages. As he waited for completion of printing, he heard the outer door to his laboratory open. A curious urge spun his chair around. His curiosity vanished, replaced by a smile.

"Hello, Clyde. Pushing your pills today?"

A loud cackled followed the question as a tall, lean man of about mid thirty came into his office toting a large sample case. Clyde Nevins was a drug representative for a huge pharmaceutical company headquartered in Boca Rotan, Florida. He parked the heavy case by the desk and sat in a side chair.

"Whew! Getting hot for November in the Big Apple," Clyde reasoned.

"Well, rest your big feet for a spell," Chip chuckled.

"Thanks buddy," he replied as he propped a hand under his chin, his elbow resting on the chair arm. "I've done some research on the questions arising from our discussion last week."

"Yeah?"

"Can't find a lead yet, but I have some advice."

Chip nodded and clicked out of the document on his computer. "Advice I can use–shoot."

"Your father had enemies in the pharmaceutical industry. Just who or where is not clear, yet. What is clear now is that you are in danger as long as you continue your father's research."

An eye arched nonchalantly. "It's money, isn't it–that old bottom line," Chip replied wagging his head.

"Yup, you got it. The drug companies are making big bucks on the family of resins directed at lowering cholesterol in America's obese population, although diabetes may be a more lucrative money maker."

He nodded, not in agreement but satisfaction, "I'm glad you can supply me with pure inositol and IP6."

"As long as it's available, buddy," he replied with a pause. "Listen Chip, and remember that these same companies exist on government research grants. If you find a cure for cancer you take money out of their pockets."

"I wouldn't call it a cure exactly; we can normalize cancer cells and reduce the size of tumors to remission, that is, in a few types of cancer."

"Well that's closer to a cure than anything I've seen. Ever since President Nixon declared war on cancer, The National Cancer Institute and the American Cancer Society choose the targets for grants and support research in its popular effort to develop "silver bullets" that statistically attack cancer cells. Biomedical scientists saw the opportunity to exist on government grants, and they synthesize these "guided missiles" with great financial success on the backs of the taxpayers. There has been very little research on understanding basic life processes—you may be one of the few."

Chip tugged at an ear lobe. "Then what is the answer, Clyde?"

He slumped back into his chair. "That's a hard question. When tax money is allocated for a new drug, no amount of convincing data can stop the money machine."

"Well, I don't intend to sit by idly," Chip confessed.

Clyde braced his hands on the desk and leaned into Chip's face. "I expected you to say that. Your research is the only viable action in America. So, my answer is to move now, move with great haste. Synthesize your cure before these money-hungry barons decide to move against you."

Clyde knew there was no reason for reminding Chip of impending danger, but he was a close friend, a fine moral man, and a good physician. Deep within the cortex of Clyde's mind he harbored a personal problem that he had wanted to share with Chip but decided instead he'd wait–Chip was too preoccupied; he saw it in his eyes.

Chapter 7

AS CHIP PUNCHED the icon on the wall by an elevator in the hospital the doors opened and revealed his mother who stood inside.

"Going my way?" she smiled.

"Hi mom," he said as he stepped inside, and the doors closed behind his entry with an audible ding. His mother had so much responsibility that he hated to pose a question, but it was too important.

"Mom, forgive me, but I have a proposal to present to the Board of Directors tomorrow morning. Wondered what you think of my chances."

She punched a logo for her floor. "You want funds for your clinical study—that right?"

He only bobbed his head, extremely interested in his mother's opinions, not wanting to interrupt her thoughts.

"You've got one powerful friend in Dr. Rosenberg. But be careful, son. I think there's a Judas on that board," she warned.

"Really?" he frowned.

As the elevator doors opened, she touched his hand with a loving smile, confident in Chip's ability. Again, his eyes painfully reminded her of how much she had loved his *real* father. If it were possible to love two men at the same time, then she had reached that pinnacle. But luckily she had Chip, a sobering reminder of both men.

"No evidence, only a tired woman's opinion," she replied. "This is where I get off," she cautioned with a raised index finger. "Remember to be careful Chip. Use your brain, son."

The young medical doctor stood spellbound as the doors closed. Her remarks stirred in his mind. Careful he wasn't, nor was his brain up to the task; he had lost too much sleep over the past weeks. Perhaps he'd finish his research early tonight and catch up on some sleep. Beverly had been very accommodating lately and he was sure she wouldn't mind if he came to her apartment early tonight.

His head shook and he suddenly realized that little Debbie needed her special treatment tonight. Oh well, he thought, not much chance of sleep tonight. The elevator stopped abruptly and the jolt cleared his mind. Chip stepped out into the basement hallway and went directly to his lab.

***

Chip sat down in his chair at the computer and clicked on a logo, then opened a file of his report to the Board of Directors. As he scanned the text he wondered if the data were adequate or even if the Board were cognizant of his work.

He slumped back into his chair and gazed into nothingness, his thoughts repaying his mother's words: a Judas on the Board. Then he thought of his father, the hours of research he'd left for him to develop into a cure. The word stuck in his throat: a cure, really, was there a chance or was it just a fantasy, a pipedream, a misuse of time. He dropped his head on the desk and closed his eyes; perhaps, he thought a few hours of sleep would refocus his mind.

No sooner had he'd closed his eyes his brain refused to sleep and he thought of Debbie's condition, her need for his formula. Adrenaline poured into his bloodstream, he stood and chastised his self-pity.

What arrogance, what nonsense! Surely there must be a cure; he had to find it–for Debbie's sake, if for no other.

He pushed back in his chair, stood, and walked to the elevator. 'As he pushed the 'up' icon, he scanned his report from memory written indelibly in his mind. The ding of the elevator broke his revelry and he stepped inside. As the elevator motors moaned, the grotesque tone buzzed inside his head. Millions of people bedridden with cancer groaning for a treatment other than radiation, someone who cared enough to break the bond of politics. Was he up to the task? Was it even possible? His father thought so, then so it was settled in his mind at least. He swallowed the thought.

Finally the doors opened and he hastened to Debbie's room. Passing by a nurse's area, he waved to the nurse he'd assigned to Debbie's case. They both entered her room.

Chapter 8

CHIP AROSE FROM the bed in a downtown apartment, his arms gyrating with agonizing sighs. He had worked most of the night and had crashed in the apartment of Beverly Banner, a receptionist at the medical center on the patient ward. They were friends, close friends, but Chip had neither made advances nor led her to think otherwise.

Beverly stood in the bathroom brushing her teeth before a round mirror. She had already dressed for her two o'clock shift at Cook Medical Patient Center. Her mind said Chip was more than a friend, but he had not realized that yet. She loved him, and that was that! There would be time for them to be alone, she reasoned. His computer was her worry, if he ever completed this research project—that was her real concern, being replaced in front of that darn computer—his mistress, but she would change that.

Since the day she had first met Chip, the day she became secretary at Cook Medical Center, the day she gazed into his handsome face, was the same day she set her mind to have him. Beverly had left a broken home in West Virginia, not her family home, but the house she had left because her boyfriend had abused her. As a naïve young and beautiful woman she had worked as a waitress in several New York restaurants until lucrative tips accumulated that paid her tuition for becoming a medical secretary.

A sudden thought preempted her revelry, and her pristine face wrinkled as she stared at her image in the mirror. She knew quite a lot about Chip's project from the hospital scuttlebutt, which had fueled her worry. *Oh God, please protect Chip. I can't bear to lose him,* she prayed.

Chip hurried and dressed behind a closed door, and paid little attention to Beverly. She was a sweet girl, a kind person, and it was awfully generous that she'd allowed him to crash each night in her townhouse; it prevented a two-hour drive to his apartment, he thought.

He grabbed an egg and ham sandwich from the table that Beverly had prepared, and quickly swallowed a glass of OJ. Too many details clouded his mind: he needed copies of his confidential handout for an extremely important meeting with the Board of Directors today. He stumbled forward while he haphazardly inserted a leg into his trousers. Suddenly wobbling, the wall by the door caught his fall, and he slipped the other leg into the trousers.

The front door slammed behind his hurried exit as he buckled his belt, and raced to his car, a briefcase dangling from one hand. The engine cranked with a twist of the ignition key, and he drove directly to Cook Medical Center as he flipped open his cellphone.

Beverly looked again in the mirror as she heard the door shut. Reality stared her in the face and questions poured from her mind: how could a busy man like Chip understand a woman's heart, asked the image in the mirror? Oh he doesn't really care for me, she realized. All he cares about is that darn research project, she thought, prompting her face to wrinkle above the bridge of her nose.

"Well I'm sick of it!" she screamed aloud, and threw her hair bush into the toilet!

Truth suddenly stabbed her in the heart; the hammer of vanity smashed her dreams on the anvil of pride. She stormed out of the bathroom, and tossed her purse on the table. Beverly made a hasty decision: Call in sick. Now that she realized the truth, she felt useless, worn out, empty—crazy! She fell across the bed sobbing as she relentlessly pounded her fists in the covers, her legs seesawing in the air.

Chapter 9

CHIP APPROACHED THE outer office of the conference room in the penthouse of the Cook Medical Center with grave doubts pounding in his mind, the seeds unwittingly planted there by his mother: a Judas on the Board. His briefcase contained a confidential handout with numbered copies for each member of the Board of Directors. Fortunately they were businessmen and medical physicians each whom had known his father. Perhaps he had a slim chance, almost invisible, but a chance nevertheless. Then questions surfaced in a moment of recurring doubts. Was his father's mysterious death, an asset, or a liability? Would the board members support his work or scoff at his data?

Questions went unanswered.

He stood facing the closed door balancing on one foot as he polished his shoes on the backside of each leg like a schoolboy on his first date. Chip pinched his cheeks to liven his face, a trick that Beverly had taught him, straightened his tie, and buttoned his sport coat. A nervous hand reached for the brass doorknob.

The secretary of the Board of Directors met him at the door in her outer office, and took his briefcase.

"Thank you, Mary—there's a handout in the briefcase."

Her smile relaxed him as she nodded. "Can I get you anything to drink, Dr. Cook?"

"Coffee, please—black."

She opened the door to the conference room and ushered Chip inside. He entered zombielike, and took the vacant seat at the end of the long conference table. Six members were seated; they waited like pallbearers, three men and three women seated on opposing sides.

"Good morning ladies and gentlemen," Chip began.

No response only gazes, an omen of rejection.

Mary sat a cup of steaming coffee and napkin by his trembling hand. She touched Chip's shoulder and whispered in his ear.

"Relax Chip; some of these board members are your dad's friends."

She waltzed by each chair and placed a package on the table before each member. Pages rustled, and then a sniff, a cough, and a grunt, all from the male board members. The ladies sat poised ready for the attack, Chip was nervously convinced.

The CEO and Chairman of the Board sat at the opposite end of the long table facing Chip. The handsome female physician focused her relentless gaze on the guest.

"Young man I am quite familiar with John Cook's pioneering work. Let me offer my condolence on your father's untimely death." She raised a gold pen to the edge of her ruby red lips. "What exactly is it you require of us, Dr. Cook?"

There it was: his opportunity, but he momentarily froze. The nights of research, sleepless nights that he suspected might choke this paramount opportunity. He sipped a swallow of coffee that thawed his throat.

"Thank you, Dr. Rosenberg. I deeply appreciate your decision to rename this facility in honor of my father. Bear with me a moment, please.

He panned the sober face of each board member like a master sergeant previewing his new recruits, as he sipped a swallow of coffee.

"My father left a wealth of raw data on the potential of inositol and its derivative IP6 as a preventive and preexistent therapy for cancer. I have thus far extracted phytate from soy beans and have isolated inositol. But I have only begun to discover the scope of its derivatives experimenting with *in viva* tests using transgenic mice." He swallowed, and panned the group. "What I need is a clinical study program on a statistically significant population of two thousand patients—"

A gruff voice like a rasp on steel interrupted. "Dr. Cook, I see on graph G6 that inositol is less effective in culture A2—what's the limiting factor?"

Chip quickly answered. "Protein, Dr. Shockley. The phosphate radicals don't assimilate as well in the presence of excess protein. Sixty percent of the inositol is excreted as waste."

The rebuttal came swiftly. "Doesn't that fact influence the viability of your clinical study?"

"Good question, Dr. Shockley. We saw the assimilation problem, and designed a treatment with pure inositol to saturate the cells, and thereafter we administered the dosage between meals, which has permitted adequate assimilation."

Silence.

Finally a second rebuttal emerged from the grim hush. "With little data on the side effects, and not complete benefits of the derivatives, isn't it difficult to ascertain what dosage to administer?"

Chip nodded. "We've conducted sufficient *in vitro* tests that prove a person of 150 pounds weight could assimilate 2800 mg per dose."

Eyes returned to the report. The only sound was the rustle of turning pages and a few grunts; even a female grunt punctuated the sound.

Finally a question broke the silence. "Is this treatment of yours really effective on pre-existing cancer cells, Dr. Cook?"

The remark implied disbelief more than sarcasm. Chip flipped the pages of the report.

"Dr. Brown, may I direct your attention to Chart C5. These plots represent tumors resulting from subcutaneous injections with cancerous mouse FSA-1. Note that metastasized cells were transported through the bloodstream into the liver. Daily injections of IP6 reduced the size of the resultant tumors. Not only reduction in size but remission of some tumors is clearly shown."

An instant rebuttal barked. "Won't the liver be overloaded?"

Chip expected the question. "The liver can be revitalized today—education is the key, here," he injected.

Another physician joined the barrage of questions. "Are cancerous cells really normalized, I mean you imply they are healthy cells after your treatments."

A hush settled over the room.

"The answer is yes, Dr. Keller. The cell phenotype is normal."

"That's astounding to say the least, Dr. Cook," Keller replied.

Chip leaned back in his chair removing a cramp in his spine. "Those are my sentiments precisely, Dr. Keller. In fact, these normalized cells have shown the ability to become productive again—even experience mitosis. Cytoplasm division is normal, and the exact chromosomes and number complement the parent."

Dr. Rosenberg sat smiling with her elbows on the table; one hand placed on her arm, the other hand's index finger lay across her powered chin. Her eyes caught the fidgetiness of one board member, not a physician who looked unconvinced. Before she spoke, the gravel voice of the old brash friend shattered the conversation.

"So, Dr. Cook you believe you can take a preexistent tumor patient who has gone through chemotherapy and radiation, and resurrect that patient back to a normal life," Harry Pennington barked, CEO of an international construction company.

Chip swallowed a gulp of coffee, every eye on his answer as he panned the group, his focus finally settled on Dr. Rosenberg's smile which somehow gave him confidence.

"When I began this research after taking over from my father, I had to devise an experiment that would ascertain if IP6 and inositol, combined or separate, could indeed prevent cancer. After weeks of experimentation, I decided to take the formula myself, even though I couldn't restrain my visions of Dr. Jekyll and Mr. Hyde.

A relaxing chuckle rumbled around the table, piercing the tension. Chip cleared his throat.

"Nonetheless, I took a heavy dosage. The tablets tasted slightly sweet because inositol has the chemical structure of sugar.

Chip's swallow was heard in the silence, every ear hung on his next words.

"Most cancerous cells, including colon, breast, prostate, lungs, and pancreas, to name a few, express a 'marker' not otherwise expressed by healthy cells: Gal-GalNac, a simple sugar. We can test the cellular contents with 80-95% sensitivity.

"Mr. Pennington, I can tell you categorically that inositol plus IP6 suppresses the marker-cell and inhibits the proliferation of cancerous cells."

Stunned gawks gripped the faces of board members. Out of the silence less technical questions were proposed by less technically trained people—financers.

"What does this program cost, Dr. Cook?"

Chip cleared his throat and took another gulp of coffee. "I estimate ten million dollars for the first year."

"Ten million," Pennington barked, and removed his stubby cigar; ashes flew into the air, and sprinkled on the table like fly ash from a coal-fired smoke stack?

The board member seated beside Pennington touched his hand as he watched the large veins swollen in his neck.

"Take it easy, Harry. Don't blow a fuse. I spend that much money quarterly in my business. The question is how much return can we expect?"

A female board member rustled in her chair, the chief accountant for a Wall Street stock company. "Ah . . . I believe the returns would be astonishing, even astronomical, considering that we would be offering a therapy for cancer without radiation or chemotherapy. But!" she paused with a raised index finger.

"But what," Pennington growled both hands laid flat on the table with his cigar gripped between grayish lips?

"*But* we cannot market the product without involving a reputable pharmaceutical manufacturing company."

Pennington removed his cigar. "We'll, hire one!"

"It's not quite that easy."

"Easy—what's so hard about signing a license agreement to manufacture?" Pennington argued.

The president of an advertising conglomerate slipped to the edge of her chair, and leaned toward Pennington.

"Harry, we aren't dealing with one of your construction projects to contract the building of a bridge in Pakistan. These pharmaceutical barons want to control all the strings—especially the strings on the purse. They are not going to allow us to market the product through their sales force without a sizeable loss of our profits."

Pennington pounded his fist on the table, cigar smoke curling in a bluish cloud toward the ornate ceiling. "Well, I have an answer for that, too. Let's form our own pharmaceutical company. We can call it Cook Pharmaceutical, Inc."

Silence persisted for a long moment of digression.

Morris McKinley, newly minted vice-president of an international bank smiled as his countenance changed to various shades of color like a chameleon. McKinley had known John Cook when he ran a small clinic in Brooklyn just after his residency at Johns Hopkins Hospital. Most of his patients were Irishmen and Italians who had little money to pay for his services. His bank had financed the building of this Medical Center when John discovered a possible cure for cancer.

"Well, since you put it that way, my bank will put up the finances. Cook Medical Center has sufficient collateral and the business plan is delightful. And I might add that I know just the company to purchase at the right price."

Dr. Rosenberg spoke. "Mary, see that the books show only the vote, not the discussion," she instructed. "And put the recorded tape in the corporate safe. Are there any more comments?"

40

Pennington could not resist the opportunity. "I have one pertinent question. What in the hell is this phytate stuff?"

Before Chip formed an answer, Dr. Rosenberg chimed in. "Harry, the term phytate is the salt form of inositol. Dr. Shamsuddin of the University of Maryland Medical Department gave Phytic acid a code name—IP6.

"IP what?"

"It stands for inositol hexaphosphate—hexa, meaning six," Rosie injected.

"Uh, six what?"

"Six phosphates groups—come into my office one day, Harry and I'll explain it on the chalkboard," she grinned.

The executive seated beside Pennington leaned into his ear. "Better leave it there, Harry."

Harry grunted, "Humph!"

Chip sat mummified absorbing the comments as he glanced at one board member seated next to Dr. Rosenberg, who had not said a word, and then the curious man gingerly rustled in his seat only to adjust his numb buttocks. Chip noticed particularly the man's body language as his eyes drifted to McKinley for some reason. His thoughts vaporized into the high-pitched sound waves of Dr. Rosenberg's voice.

"Before we vote," Dr. Rosenberg advised, "I suggest we take ten to stretch and gather our thoughts. Chip, I think it best if you step outside. Thank you for your excellent presentation," she smiled.

The board members gathered around a sofa and table at the opposite end of the room and sipped on wine. Chip rose from his seat and stepped into the outside office, leaned against the wall, because he'd lost support of his weak, trembling knees.

Although time had elapsed only a few moments, it seemed an eternity, the relentless crawl of time. The sudden buzz of his cellphone sent his heart leaping into his throat.

"Hel . . . hello."

"Chip, this is Clyde. What did the board say?"

He relaxed. "They're voting now."

"Then you have a chance. Listen, I've run upon some information from a buddy in a rival pharmaceutical company. Are you available for lunch tomorrow?"

Chip straightened his torso against the wall, blood surging into his legs. "You bet. I may have a person for you to check out . . . wait Mary is calling me back. Talk to you tomorrow, Clyde."

Mary beckoned with an encouraging smile on her face. She was extremely fond of Chip, having known him since he was a toddler. She was a close friend of his mother, Jenny Lynn, and idolized his father, Dr. John Cook.

Chip entered the chamber behind Mary, surprised by a beaming smile on the face of every board member except the silent one with the curious body language.

"Well, young man," Dr. Rosenberg greeted, "You bought the bank. You have the ten million and we may soon be the proud owners of a subsidiary to manufacture your product."

Chip's knees buckled and he fell into a chair deflated.

As the board members stood chatting, Dr. Rosenberg strolled over to Chip's seat and sat down beside him. Her face sparkled as if she were twenty again.

"Chip, my boy, you gave a blue-ribbon presentation. You're the spitting image of your father. I thought a lot of John Cook, you know. I knew him before Jenny Lynn came to New York. He helped me not once, but several times. He was directly responsible for bringing me here from Johns Hopkins.

He played no small roll in my present position here at the Medical Center. I owe him, and now I can repay through you. Don't let your father down, Chip."

Chip swallowed. "Thank you, Dr. Rosenberg."

"Rosie will do, Chip. That's what your father called me."

"Then thank you . . . ah, Rosie. Thank you indeed."

"Incidentally," she smiled. "Johns Hopkins can provide maybe a hundred bed patients and significant walk-ins. And one thing more, I know a three-year resident at Johns Hopkins who could use the experience. Her name is Dr. Marcy Curtis."

"I can use all the help you can provide . . . really don't know how to thank you . . . Rosie."

"Seeing you launch out on this groundbreaking venture warms my heart, and that's enough thanks for me. You find that cure Chip, and you can purchase Cook Medical Center," she smiled, a glow of confidence on her face.

As he assimilated her remarks he caught Mary's glance, excused himself, and strode to the entry door. Mary met him in her office and asked him to take a seat as she closed the door behind his entrance.

She walked around her desk and sat down, then clasped her hands together finger-to-finger. Chip glanced at her polished nails and lifted his eyes; his curiosity was on the verge of exploding.

"That man you were gazing at is Dr. Kosaku of Gunma University, Japan. He owns a small drug representative company based in Boca Raton, Florida. The company is rather insignificant to his prestigious university professorship, but the Japanese are big on owning US companies."

"Hum. Why would he object to more sales of IP6?"

43

Mary expected the question. "The Pharmaceutical industry has a strangle hold on what drugs reach the market. At present they are heavily loaded with windfall profits from social drugs. Dr. Kosaku cannot afford to lose 'a bird in hand' for 'one in the bush,' so to speak."

"You speak very well."

Questions surfaced, but answers were few—only Clyde's warning and his mother's suspicion confirmed by Mary's information. The muddiness of his thoughts was somewhat clearer now. Mary was correct: This Dr. Kosaku deserved his attention. Yet there wasn't much time to spare. And then Clyde's warning surged to the pinnacle of his thoughts: "There are those willing to pay any price for your research, even murder."

Chapter 10

THE SUN DAWNED over New York City and the celestial light reflected from the windows of the tall buildings. The nighttime traffic announced breaking day with honking horns and fatigued drivers headed home after the nightshift. The sun promised a cloudless day, but a young physician at Cook Medical Center had other things on his mind besides his research.

Chip Cook sat at his computer, and closed his cellphone after a conversation with Clyde Nevins who was in town following up on their lunch appointment. He had promised to meet him at a grill about a mile from his laboratory. He wasn't achieving much at his computer anyway; his mind was a hundred miles away, and he anticipated that a talk with Clyde would be beneficial to his emotions and to the research.

All night he had thought over the meeting with the Board of Directors of Cook Medical Center—his mistakes, his victories, but mostly the quiet board member with the suspicious body language, Dr. Kosaku. But his dreams drifted in-and-out of Rosie's offer that sent a resident physician to help in the clinical study . . . oh what was her name, Marcy something . . . Curtis, Dr. Marcy Curtis, he thought.

He shut down the computer after saving his work on nude mice *in vitro* study with type L2 cancer that he started two weeks ago. He had selected several tracks on immunity therapy. He was far along on a test with monoclonal antibodies (MABS) used to target cancer cells selectively. He had also selected two avenues: interferon, specifically mouse interferon that had been affective against mouse leukemia.

The second *in vitro* study was with interleukins, mostly focusing on IL2, which had shown the ability to kill tumor cells in mice. He was at the stage of introducing the IP6 derivative singly and in combination with inositol. The early results were astounding: the tumors were shrinking, to his delight! Increasing evidence of cell normalization was especially significant and gratifying—exactly what he had told the Board of Directors.

As he moved toward the door a sudden series of thoughts overwhelmed his mind. Out of the synaptic juggle came the image of Dr. Kosaku. Was he the Judas? Maybe Clyde knew something about the small drug rep company in Boca Raton that he had emailed.

The parking lot beneath the building was cold and breezy and Chip lifted the collars of his knit jacket. He beeped his car with a press on his key, and he saw the headlights blink near the far wall. As he strolled to the car, he punched a code in the door keypad and the door unlocked with a 'click'. He stepped inside, and just before he closed the door, he heard a loud voice screaming at the stairs door. He jumped out and ran toward the building janitor, who stood waving his hands and screaming something. They met in the first row of cars.

"What on earth is it, Sam?" he asked worriedly.

"I thought I'd never reach you. Dr. Cook!" he gasped.

"What is it, man?"

He placed his hands on his knees huffing and puffing for air. "I saw a suspicious character doing something under your hood."

"Really!"

46

"I'm so rattled I can hardly think. Please check it out sir—for my sanity," he pleaded leaning on a mop handle as he wiped the perspiration at the nap of his neck with a dusting cloth.

"You bet we will."

The curious couple went back to the car, and Chip quickly opened the hood. What he saw was a myriad of parts and wires crammed around the 6-cylinder engine. But there on the firewall he focused on an unfamiliar package. It had output wires; two, one hooked to the battery, and one twisted around a screw on the ignition switch cylinder. They both rationalized that it certainly wasn't a part of the car.

Chip rapidly flipped open his cellphone and punched a memory key. "Clyde . . . Where are you now . . . I see. Could you swing by my parking garage, I need a ride . . . thanks, buddy."

He closed the phone inhaling a deep breath, and then it hit him. Somebody wanted him dead!

He and Sam both nervously laughed with release of anxiety. The sudden squeal of tires turned both heads. It was Clyde's car pulling into the parking lot. The car stopped beside Chip's car.

"What gives here?" he quizzed, stepping from his sedan.

"Maybe you can tell us—take a look under my hood," Chip explained.

Clyde gazed into the motor-well and immediately saw what shouldn't be there.

"It's C-4—military issue. I'd say this is the work of a hit man."

"Whee!" Chip gulped. "Okay, let's talk about this over lunch—Sam here's a hundred dollars. Take your wife out to dinner. And thanks my friend, you saved my life," he said stuffing his wallet back into his rear pocket.

Chapter 11

AS TWO FRIEINDS from college days drove down the boulevard toward a grill for breakfast, Chip alerted the police department, and placed his cellphone in his inside coat pocket. Clyde waited silently for Chip's report as he steered through the morning traffic.

"They're sending a bomb squad to remove that package. I'll have to go down to the Midtown Precinct and make a statement after lunch."

Clyde had something serious on his mind and didn't join in the chitchat. He finally swung into the parking lot of the grill and parked. The two men walked through the door, and Chip finally ordered a table. A waitress led them to a quiet place in the next room, and they settled into their chairs, the atmosphere quiet as a tomb. While they mused through the menu, the waitress returned to take their orders.

As she left with their selections, Clyde fidgeted with his fork. His elbow propped on the table with his head resting in his hand. The other hand scribed parallel lines through his napkin with his fork. It was a bit unnerving for Chip, not the lines but the screeching noise of the fork tines. Something was bothering Clyde; it wasn't his normal demeanor, nor was it the bomb placed in his car—that was another matter for the police.

"How are Margaret and the girls, Clyde?"

He dropped the fork and clasped a hand over his mouth beneath his nose. His eyes squinted as if a knife had been plunged into his gut.

"Chip, there's something I've wanted to discuss with you, personal," he said fingering the fork.

"What is it Clyde, are you sick?"

Clyde moved a glass of water aside, rocked back in his chair; his hands were an animation of nerves. "It's little Debbie, Chip. She has leukemia."

Little Debbie, as the family lovingly called her, was only twelve, the youngest of a female trio: Debbie, a prodigy violinist, Simon age fourteen, a virtuoso cellists, and Clara age sixteen, classical pianist. Their Mother, Margaret taught piano at home.

"Oh, Clyde I'm so sorry, really I didn't know."

"What I mean to say is . . ." he dropped the fork. "Would you consider administering your formula to her?" he carelessly blurted out.

Chip sat silently for a brief moment. "I would like to put her in my clinical study group for testing in about six weeks."

Clyde regained his composure. "Chip, my old friend, I'm not sure she has six weeks."

Silence of empathy persisted.

"You are aware of the risks, and Margaret, too, I trust? The formula has not been tested clinically, yet."

Clyde's head bobbed decisively. "Six weeks from now we will place her in your clinical study group, but right now she must have your treatment if she is to survive."

Chip exhaled a deep sigh as his Hippocratic Oath flashed across the retina. Morality seized his thoughts. A series of oxymoron's floated in the cortex of his mind; rules, law, loss of his license. And then his mother's compassion registered in his genes, and formed his final response.

"Okay, Clyde. I'll do it for little Debbie."

Clyde's face flashed alive, "That is just super—when?"

"Tomorrow, but I'll have to go to the lab and prepare the dosage. I want you to call Margaret right now and start Debbie on a fast: no more solid food today or tonight until she arrives at the medical center."

"I see." He gazed into his friend's hazel eyes. "Chip, I trust you with my child more than I trust the bomb squad about to remove that C-4 from your car."

Chip grinned embarrassingly. He puzzled over the toxicity of the formula if administered to a small child, but Clyde had something else to say.

"After we talked briefly yesterday, I had lunch with an acquaintance, a rep in another company out of Boca Rotan. He tells me that the owner of the firm has contracted to build a synthesis laboratory on site."

"A manufacturing facility, huh?"

"Is it important?"

"Could be if they manufacture pharmaceuticals."

"What are you talking about?"

"Don't know exactly," Chip replied as he reached over and placed his hand on Clyde's shoulder. "Clyde, I really think it would be much safer and legal if you register Debbie at the Medical Center tonight—let's not wait till tomorrow. That way she can have better care and we can start the treatment intravenously."

A broad smile of relief spread across Clyde's tanned face. "Margaret would like that—what time?"

"I'll leave admitting instructions with the nurse and you check in Debbie about nine o'clock tonight."

Clyde's face revealed embarrassment. "I don't know how I can ever repay you, Chip."

Chip took his best friend's hand. "Let's hope Debbie responds positively to the treatment—that's more than sufficient payment, Clyde."

Chapter 12

"SO YOU'RE THE son of the famous Dr. John Cook?" the police said, reading the form in his hand while rubbing the unshaven bristles of his five o'clock shadow.

"It's no secret," Chip barked.

An eyebrow arched. "Just listen for a moment Dr. Cook. A bomb placed in your car is a serious matter," he remarked moving around his desk at the NYPD Midtown precinct. He threw one leg over the corner of the desk; one hand gripped his chin, the other hand in the opposite armpit.

"Forensics has traced that C-4 to Fort Dix. It's a federal case now, but I'm going to put a guard at your lab until the federal boys arrive."

Chip loosened his tie. "Is that necessary?"

"Necessary or not, until I know what's going on here, I'm taking no chances," he barked.

Chip was too smart to react physically. "Well, place your guard or not, I'm going back to the lab when we finish here and work a while."

The slid his leg off the desk and slumped back in his chair. "Do what you must, but don't take unnecessary chances, young man."

Chip stood. "I don't need a nursemaid, detective."

Henson rose from his seat, both hands on his hips. "Why you arrogant—listen, doctor. I could hold you here for 48 hours."

Chip tightened his tie.

"When you decide, you can find me at the Medical Center until late tonight."

Henson pushed his chair aside, aware he had an intellectual on his hands, so he changed the subject.

"Who is this close friend of yours, a drug rep, I believe."

"Clyde Nevins? Why he's no suspect," Chip angrily replied, as he pulled out his cellphone and flipped it open. "Here's his number," he said, and held it in his face.

The detective jotted the number on a pad. "We'll put a tail on him, too."

"Maybe you didn't hear me—Clyde is not a suspect," he demanded.

Henson held his cool; he'd worked cases like this before, witnesses who had all the answers and no real clues.

"Whoever put that bomb in your car is probably watching you both right now, Doctor."

Henson's logic finally seized Chip's emotions, and he suddenly realized how wrong he'd been. All he wanted was instant answers, the priority of a physician. Yet, it occurred to him that he'd spent months, even years, without the right answer. He knew nothing about justice, only that his father had not received it. Suddenly he blushed in a cowardly manner as he sheepishly bowed head.

"I'm sorry, Detective Henson. This kind of thing is over my head," Chip replied remorsefully.

Henson smiled as he poured two cups of coffee, and sat the decanter on its base. "Sit down, Dr. Cook. It would help if we knew the motive for this attempted hit."

Chip pulled back a chair and sat. "It's a long story," he sighed.

"I've got until six. Maybe this cup of coffee will help you think," he replied, and handed over a cup, while he balanced another hot cup in the other hand. Henson had changed his first impression of this young man. He must be under a great deal of pressure, and it was his job to get answers.

Chip gripped the hot coffee cup, leaned back in his chair, and crossed his legs. He wrapped his hands around the soothing warmth of the cup.

"Where can we begin?"

"At the beginning," Henson advised.

Chip took a sip of coffee, and sat the cup on the desk, his mind sorting facts, people, and places.

"My father died when I was in medical college. My mother firmly believes that he was murdered."

Henson released his cup. "I've heard rumors. It could be the same reason someone wants to kill you?"

Chip decided then that Henson was closer to the case, and deserved his cooperation. "That bomb was a rude awakening, I must admit."

"Let's cut to the source," Henson advised sipping from his hot cup.

Chip deeply sighed. "The source is the IP6 Code," he replied flatly.

Henson's eyes popped open. "Excuse me."

"IP6 is a potential preventive for cancer—"

"Did you mean it *cures* cancer?" Henson gasped as he spilled coffee down the front of his shirt, and flooded his freshly cleaned trousers.

Chip smiled, not at the spill, but his reaction. "It's a therapeutic treatment for preexisting cancer without chemotherapy or radiation; in fact, it normalizes selective cancer cells."

He coughed, spewed coffee over the front of his wet shirt. "Well I can think of a number of motives here," he replied soaking a handful of napkins over his trousers.

"But why do they want to kill me?" Chip barked.

"It's your formula they want. They think they must kill you to get it and keep it," he replied drying the liquid on his shirt with a second handful of napkins.

Chip gulped. "But murder."

Henson's face smirked. "It happens every day for less reason. In your case, every pharmaceutical company from here to Japan would like to get their hands on your research. Doctor, you have something worth murder!" he exclaimed.

That remark stuck in Chip's throat. "I might have something else of value, too," he ventured.

Henson crossed his legs and felt the liquid trickle into his shorts. "That's more like it."

"I'm not at all sure there is a connection, maybe only circumstances, but there is a small pharmaceutical rep company in Boca Rotan, Florida that has something fishy going on."

"Explain "fishy."

"They're building a synthesis laboratory on the site."

"Is that unusual?"

"Drug rep facilities don't normally branch into manufacturing drugs; they represent a number of manufacturers, and serve their clients without bias."

"But some larger companies do—why fishy?"

Chip only nodded, reviewing his statement as to whether it were fishy or just smelled bad to the detective. He rephrased his response.

"The Board of Directors of Cook Medical Center has just awarded me a grant for clinical studies, and Dr. Kosaku, a board member, walked off with a confidential copy of my research notes on the IP6 Code."

Henson forgot his wet shorts. "I see, so you think this Dr. Kosaku is about to steal your research."

54

"There could be a conspiracy here—that's what I think."

"You let me decide that, doctor. Have you got anything else to go on?"

"Yes, I think so."

He frowned. "Well speak up man!"

Chip gripped his bottom lip between thumb and index finger recalling Clyde's recent cellphone call. "Dr. Kosaku owns this medical rep facility now under expansion as a manufacturing facility."

The detective dropped his glance, and marched around the desk with his forehead wrinkled, his mind busily sorting facts, his trousers without crease. "Well, maybe we *can* say your story is a bit fishy," he said. He raised his head, a sparkle in his eyes. "Perhaps we can set a hook."

Chapter 13

CHIP'S LABORATORY LOOMED dark except for the safety lights in the hallway. He scrambled in his pocket for the door key, inserted it into the lock, and pushed open the door. The young physician raced to the vault as he pulled out his code card. The vault was a bank-style, walk-in safe that his father had contracted construction sometime after they returned from Africa. When Chip had finished his residency and took over the research, he had modernized the lock with a keypad security system that required a coded card for access. Chip inserted the card into the pad by the door, and a green light lit. He wrapped an arm around the locking wheel and rotated it until the huge steel bars retracted into the massive door. With all his one-hundred eighty pounds against the door, plus a push of his legs, the door finally swung open. Chip entered the vault and walked by the biohazard containers while he pulled out his key ring. Because of the many keys on his ring, he had color-coded several important keys; the red key was the one that opened a tray. Finally he inserted the red key into one of the trays fashioned like a safety deposit box at any bank. He pulled out the tray and retrieved a sterile bag of serum. The label was marked: Inositol and IP6 50:50 solution.

Quickly he placed the bag inside a stainless steel container with a handle that looked much like one of those gallon-sized paint cans. He rushed to the entrance, sat the canister on the outside floor, and swung the door closed. Carefully he placed the steel container on a table by his computer, and returned to the walk-in vault. He rotated the wheel that engaged the steel bars until the green light on the pad blinked "off" and a red light blinked "on," indication that the vault was locked.

Chip grabbed the canister by its handle and hurried to the laboratory door. Again he rushed to the elevator which took him from his basement lab to the lobby of the patient wing. The elevator doors finally opened and Chip stepped inside just as the doors closed.

A person dressed in black tights and a dark raincoat exited the stairway door into the hallway where he had hid until the physician passed. He had watched the elevator until the doors closed behind Dr. Chip Cook. He ran to the lab door and inserted a lock-pick, and carefully clicked through the tumblers. The door finally unlocked and he slipped inside and closed the door behind his entry. The intruder followed the beam of a penlight to the vault and inspected the locking mechanism. Even with his professional experience, it was too sophisticated even with a cutting torch. It was constructed of a new metal series of hardened steel developed for the Army for the M64 tanks.

His information was wrong.

In desperation and increasing anger he beat his gloved fist on the door and kicked over a trashcan while mumbling profanity—his cousin was an idiot. The door required a special card that triggered an electronic key, without this card entry was virtually impossible. He deeply sighed and glided to the computer as he quickly retrieved a square plastic container from an inside pocket. He took a seat at the computer and pushed the power button. He briefly scanned the icons as he removed a CD from its plastic case, shoved it into a slot, and clicked the mouse on the icon labeled "computer." A screen flashed on the monitor, and he clicked on "directory file" and burned the entire file onto the blank CD. When the coping was complete, he closed the file and shutdown the computer gloating as he removed the CD. Now rationalizing that he had the files of Dr. John Cook, the CD went into its protective case, and snugly into his inside coat pocket.

His hand patted the bulge on his chest grossly grinning. At least he had the files, if not the formula. The consequences went to his stupid cousin.

***

Chip left the elevator on the third floor reserved for the cancer cases, and walked toward the nurse's station, in his hand swung a steel canister. He spoke to the floor nurse and discovered that Debbie had already transferred to a room, and her medical records had also arrived. However, the lab tests he had requested where not yet completed. The nurse presented a note from Debbie's mother that said Debbie had been fasting for ten hours before she entered the hospital; that made a total time of fourteen hours, sufficient time for beginning the test.

He opened the steel canister and took out a plastic hang-bag, as a nurse left Debbie's room and walked toward him. He gave her instructions to set a drip rate of 3-second intervals for 10 minutes, then to slow the rate to 5-second intervals. She was to stop the procedure in one hour. The remaining formula was to be placed back in the steel canister and into the office safe until Chip transferred it back to the vault.

The nurse bobbed her head as she took the formula bag. "Yes, Dr. Cook, it shall be done." She snapped a finger at an orderly, who followed her with a stand, and they marched into the Debbie's room.

Chip returned to the nurse's station and found a coffee pot. He poured a hot cup, and noticed the nurse had flagged him.

"Dr. Cook the lab results are here," she replied, and handed over the file.

He quickly scanned the lab data: It was leukemia, an aggressive variety. Debbie's pulse rate was weak because she had had her last radiation treatment in just over ten days. However, her blood pressure was excellent, and no surgery was planned in this initial testing. He made entries on the form that requested 12-hour blood tests with the results networked to his basement laboratory computer. He gave the file to the nurse with some instructive conversation, and strolled off to Debbie's room.

The attending nurse already had the formula on-hook with the feed line fixed to a 'Y' cannula taped to Debbie's wrist. Drip rate was set for the ten-minute sequence with automatic changeover. He smiled at Debbie lying in the bed, her two sisters stood at the bed's end.

"Hi, honey. How are you feeling?"

She forced a smile. "I'm fine, Dr. Cook—just a little hungry."

Chip took her hand as released the chart hung on the end of her bed, noticing that she had no fever. "Well, in about an hour you can eat whatever you like."

That announcement brought a smile to Debbie's sour face. The bright smile brought tears in both her mom and dad's eyes; although reserved, there was still hope in their tired faces. Chip beckoned to Margaret and Clyde who followed him outside the room, and closed the door. Debbie's two sisters waited by the bedside.

Margaret stood petrified as she leaned against the wall. Clyde seemed hopeful with his arm around his wife's waist. Chip gave them his best bedside manner out in the hall.

"Debbie has non-typical hairy-cell leukemia– that's why I think the interferon was only partially effective. Her vital signs are good. I don't expect any toxicity to develop. There is nothing you can do for the next twelve hours when we will take a blood test. I want to be sure the inositol has time to saturate the cells without a change in osmotic pressure. If you want to stay, we can arrange a cot for you in the waiting room. That's it for now. I would advise that you and the girls get some rest."

Margaret extended her arms and clasped both of Chips hands, a tear streaming down her cheek.

"Dr. Cook, I know your formula is untested and you are taking a great risk, but it's Debbie's last hope."

Her face suddenly flooded with tears and she threw her arms around Chip sobbing. Chip allowed her to cry, the best therapy he could prescribe. Finally Clyde gingerly took Margaret's arm.

"It's okay, honey. Chip is the best doctor in the country for this treatment. I trust him implicitly."

Chip placed a hand on Clyde's shoulder: no words were needed. Clyde decided he'd take the girls home, but Margaret wanted to stay. Chip agreed and gave his departing thoughts.

He finally closed the door and walked away toward the nurse station. When he reached the station, he checked in with the floor nurse. She confirmed his instructions on safeguarding the formula and the twelve-hour blood test. He decided he'd retrieve the formula in the morning, right now he was going: where? Well, to Beverly's townhouse, where else, he thought.

***

A man wearing a raincoat stood at the entrance way to the basement lab. In his hand he held a plastic case containing a CD as he waited for his cellphone to connect the number he had punched. While he waited he considered what must be done to fulfill the task he was instructed by his cousin. The phone finally connected.

"Yes," said a nervous voice.

"The safe cannot be opened, it requires a special card to trigger electronics," he advised.

A moment of strange silence hummed over the line.

"Did you get the other information?"

"Yes."

"Good," replied a more confident voice. "Deliver it to the dealer."

"Understood—but don't call me anymore—I'm through with you guys."

Chapter 14

A TALL, LEAN AND handsome man, a radiographic salesman, entered the front doors of the Cook Medical Center rolling his sample kit, and took an elevator to the third floor. The spacious lobby eventually led to the radiology imaging area. He was familiar with the radiology wing because he'd eaten in the cafeteria before, which was on this floor down the hall from the CT installation. He had supplied radiographic plates for this unit on a few occasions. And he was ready this afternoon to offer an enticing bonus just to re-establish an account. He'd lost touch with the management and knew neither the person in charge, nor even who might be the buyer of his goods. But he may luck up this day—the byline of all salesmen.

Several nurses and technicians whizzed by him in the hallway and two physicians streaked passed, but none he knew personally. As he sneaked about in his meddlesome manner, he finally turned at the intersection and followed the overhead signs to the CT imaging area.

A woman exited a door in the hallway up ahead and he thought he recognized her face, but she spoke first.

"Why Brian—it's nice to see you again."

He hardly believed his eyes.

"Well this is an honor, Jenny Lynn."

He had been acquainted with this Training Director of nursing procedures since he was once in training at this very medical center. She had taught a course in the proper procedure for locating veins in elderly patients, how to insert and inject without damaging tissue, all with the least pain for the patient. He could hear her lectures now. After it became clear to him that he had neither the brains nor the aptitude to be a medical doctor, his uncle, a respected physician, advised him to pursue pharmacy. Even pharmacy had not interested him, and he opted for a technical salesman position, sales in processor films for imaging machines, including a line of radiographic film plates. His uncle added insult to injure when he refused to use his influence in the pursuit as a radiologist degree, and he had dropped out of his first year of medical school under the added pressure of extreme headaches. He enlisted in the Army for a three-year stretch that ended in two years with a medical discharge, again for extreme headaches that put his buddies at risk.

Jenny Lynn Cook had been a kinder influence in his life and he had somehow been drawn by her kindness, a kindness that had evaded his youth when his mother had died of cancer while he was just a teenager. Jenny Lynn had a similar situation with her son, Chip, and that is what attracted them both, he thought.

Chip Cook had breezed through medical school and Brian had been jealous of his accomplishments and his family relationship with such a kind mother. The Chief of Residency was responsible for the final decision to fail him out of medical school, and he suspected that was at the urging of his physician uncle. That event had crushed his ego, and it had taken some years of recovery; at times under a doctor's care, to curb is psychotic, revengeful compulsions. It was only because of Jenny Lynn's motherly instincts that he had decided to seek other employment.

"Seems you have been successful in your chosen profession, Brian," Jenny Lynn reasoned.

He parked the roller kit. "It keeps me outside and I like to travel."

"That's great—medicine is not for everyone, you know. You look so well," she responded as she extended her hand.

He took her hand with a respectful smile, and the woman he most revered walk away.

He bowed his head. Somehow he felt dirty, filthy even, and that only because he had stood for a brief moment in the presence of innocence. She was right, as always, medicine was not for everyone. He had planted the seeds of evil desire, and had reaped a neurosis that drove him to the edge of insanity, unless he had the medication he carried in a small pill bottle. Otherwise he was content, successful, but lonely, always alone, and painfully aware that his condition was often spontaneous; the medicine was not a cure, and he never knew when a spell was near until it happened.

He gathered his emotional strength and entered the cafeteria; he needed something stronger than coffee, but caffeine had its use at the moment. While drinking coffee he made his plans for the evening.

Chapter 15

A DEPRESSED AND rejected woman of noticeable beauty sat on a barstool at an all-night pub across from the Cook Medical Patient Center in the late afternoon. She swilled down her last two glasses of scotch and soda, and had successfully drowned her many troubles in the wasteland of alcohol addiction. Several men glanced at the pathetic woman for only a moment, and realized that she was an alcoholic. But some of these men were alcoholics, too, and they felt her pain. It was obvious that she was too drunk for meaningful conversation—not worthy of wasting money for purchase of another drink. No one knew that Beverly Barrow held her liquor like any beer-belly man, but not her composure.

A clean-cut and well-dressed man with a trimmed beard and manicured nails accented by Calvin Klein "Obsession" cologne slid on a stool beside her. Beverly quickly caught the smell of his cologne and slowly turned into the face of a handsome man.

"What's own your mind, buster?" she barked, wobbly.

There was no response to her comment; instead, he ordered a drink.

"I'll have whatever she's drinking, please," he told the bartender.

Beverly pushed her empty glass forward. "Give me another round," she mumbled.

The man placed his hand over her glass. "She has had enough—I'll take the bill."

Beverly cocked her head. "Are you my nursemaid?"

He smiled. "I'm your protector—will this cover the tab," he said to the bartender as he tossed a fifty-dollar bill on the counter.

"Yes, sir—with change."

"You keep the change for your trouble."

Beverly managed a smile. "Big spender, huh?"

He took her arm. "Let me take you home."

An eyebrow arched without objection as she swilled down his drink. "Why not?"

The man smiled as he assisted her from the stool, and placed her coat around her shoulders. He led her out of the bar and toward his car parked on the curb. The man surveyed the street in both directions, and opened the passenger door. It took a moment for Beverly to navigate her long legs into the sedan. Before he slammed the door closed, he pushed in her coat.

The Lincoln pulled away from the curb and motored around the block. Beverly snuggled up to the man's side lonely for attention that Chip Cook had never given her, even after she had done everything but rape him.

"You got a name, Sir Galahad?"

He chuckled with a disarming smile. "Let's not be too personable."

She gripped his shoulder. "Where have you been all my life?"

He released her grip. "I just arrived yesterday. I've noticed your frequent trips to that bar."

"Are you lost?"

Again he chuckled. "I'm a salesman from Rochester—make my calls over here about twice a week."

She snuggled closer, eyes closed.

***

A Lincoln sedan finally pulled against the curb of a townhouse and stopped, the headlights dimmed off. The driver gripped Beverly's shoulders and gently shook her awake.

"This is your place, isn't it?"

Her eyes squinted as she panned the row of houses. "Yeah, this is home. Want to come in for a nightcap?" she suggested, hair caught in her lips.

He gazed into her eyes with a manly grin that disguised what he really thought. "No, not tonight, you need rest if you're going to keep your job."

She managed a grin as she fuzzily thought: Oh, that's so sweet. Chip Cook never was so thoughtful or kind. One day she'd fix him.

"What are you thinking?" he asked.

"Oh, nothing—walk me to the door, would you?"

His eyes said 'yes,' and he opened the driver's side door and walked around the hood. Beverly watched his every move: *My, he's tall and handsome.*

He opened the passenger side and ushered Beverly out. She suddenly realized that her legs refused the command of her brain. The young man noticed her demise, and slid his arm around her waist, steadied her stance, and they slowly walked the short distance to the four steps that rose to her porch. Beverly had very little trouble navigating the steps; in fact, the night air had cleared her head somewhat. As they reached the porch, Beverly turned in his arms face-to-face.

They stood in a silent stare.

He kissed her gently.

She reciprocated.

He took the key from her hand and inserted it into the lock, kicked opened the door.

"Maybe I will have that nightcap, after all," he said slyly.

He took her into his arms and carried her inside. The door closed.

***

Morning sunlight shafted across Beverly's bed and lit her smiling face. She fingered the curlers in her hair, blinked twice, and then stretched her arms skyward. The memory of last night spread a pleasant expression over her face. Was it a dream? Her arm reached across the other side of the queen-size bed and squinted at the clock. A frown; she bounced upright in the bed, quickly propping on her elbow as she discovered a note pinned to the pillow on the empty bedside: *Beverly, you have a good day. I'll be back in town this Saturday—meet you at the bar, same time. Mr. X.*

She bounded out of bed in a short nightgown giddy as a schoolgirl. The closet doors flung open, and she pulled out a clean outfit and laid it on the bed. She went to the kitchen and poured a glass of instant breakfast to which she added a raw egg. She waltzed to the bathroom, humming a ridiculous tune from college days, twisted open the tub facets, and tested the temperature with one hand. She raced back to the kitchen as she swilled down the chocolate drink.

Beverly checked the time, hurried to the bathroom and disrobed, then quickly stepped into the tub. The humming returned with words, just as ridiculous as the tune. Her thoughts were not about whether she was late for work or even if Chip Cook cared, only about Mr. X, and their next meeting on Saturday night. Gosh, what day it is, she thought: *Monday, oh how I hate Monday's.* But not anymore, Mr. X had seen to that—what a man, she thought.

Chapter 16

THE RESEARCH WING of Cook Medical Center was alive with nurses and aides busily registering a sizeable group of applicants who had volunteered as patients in a clinical study that featured a new cancer treatment. The many applicants were streaming in from all parts of the East and the total group could exceed the two thousand-target numbers for a clinical test of this magnitude.

Dr. Chip Cook walked to the nurse station with a clipboard under his arm, his mind triggering one question after another.

"How many applicants so far, Julie?"

The head nurse assigned for the clinical study scrolled the computer monitor.

"I'd say about three hundred bed patients, maybe a thousand walk-ins," Julie replied.

Julie Martinez was a close friend of Dr. Cook's mother and she had spent many nights on the patient ward under Jenny Lynn's supervisory. Last summer she had attended some refresher classes at Johns Hopkins with Dr. Marcy Curtis. Marcy had seemed like a nice kid, and that's what she was, bright, yet she thought, too young for a third year of residency.

"I see, well let's keep registration open. How many beds do we have available on this ward?" Chip advised.

Julie stirred from her thoughts. "The administrator gave us three hundred fifty beds, and I think that's adequate with the large number of walk-in patients available."

"Good. That means we'll need probably another seventeen hundred walk-in patients for a statistically significant program."

"Oh, yes—almost forgot. Johns Hopkins sent us five hundred applicants for walk-ins and promised more," she replied jotting a note on her pad.

"That's super, Julie, oh, and would you see that Dr. Curtis contacts me as soon as she arrives?"

"Can do, Dr. Cook—oh, I simply forget, Dr. Curtis is waiting in your office—arrived fifteen minutes ago."

"Yes, well would you tell her—scratch that, I'll see her after my rounds?"

Chip left the ward and took the elevator to the cancer wing expecting to check on Debbie while going on to his lab. He had authorized another treatment this weekend and reminded himself to talk with his research laboratory assistant; he needed his assistance in the CT (Cat) scans.

As the doors opened on the patient wing, he spotted his assistant who had dropped by the storeroom for more supplies in his basement lab. He waved and called.

"Kyle—could I have a word?"

"Afternoon, Dr. Cook. The blood samples you ordered for Debbie Nevins are ready."

"Thanks, Kyle. Listen, I'm going to need your assistance on some CT scans in the clinical study, and you'll be doing double time in my lab preparing dosages. You might want to reschedule your time for medical classes. I'll speak to you supervisor, if needed."

"That won't be necessary, Dr. Cook. I work best under pressure. And I'm anxious to learn all I can."

He nodded his head. "You'll make a fine physician, Kyle. Keep up your good work," he replied, and his mind suddenly flashed back twenty years when his father had made a similar remark during the Africa trip? Somehow he had had several flashbacks lately. Oh, well, he thought.

Kyle wheeled the cart of supplies to the elevator and Chip dropped by the nurse station and found Debbie's file. He never interrupted his stride to the patient's door as he read the reports. Glancing up, he pushed open the door, and found Margaret seated by the bed reading a magazine.

"Oh, Dr. Cook!" Margaret exclaimed.

He smiled and lifted several pages in the file. "How are we feeling tonight, Debbie?"

Debbie gave him her best effort and forced a weak smile. "The ice cream was wonderful.'

Chip fingered the lines feeding into her arm. "This glucose should give you some strength." He moved to the monitor that recorded in real time her statistics. "How would you like a full meal tonight?"

A weak smile silently voiced her approval without much effort. "Steak!"

"Chicken, I'm afraid," he smiled.

Margaret stood and dropped the magazine in the chair. "Chicken will do just fine."

Chip gestured Margaret to come outside with a wag of his head. He turned to Debbie and touched her tiny hand.

"We'll be moving you into the clinical study ward soon. You'll get constant care during the study. Your tests are doing better than expected."

Margaret leaned against the outside wall beside the door with her arms folded over her chest. Chip closed the door behind him.

"How is she, really, Dr. Cook?"

71

He sighed deeply. "It's too early to really tell. I've scheduled another treatment for the weekend. One good sign: Her blood samples show good levels of white and red cells. Most importantly, the cells contain high levels of inositol."

She frowned. "What does that mean?"

He smiled. "It means the inositol has evaded the protein interaction in significant quantity. Really, Margaret–Debbie is going to improve, I promise."

Margaret's hands covered her mouth; tears flooded her tired eyes. She cried profusely.

Chapter 17

CHIP COOK FINALLY reached his office much later than anticipated, all the while worried that he'd made a poor impression on Dr. Curtis by his lack of punctuality. Yet, when he entered his office, he found her seated at his desk reading Debbie's file.

"I'm sorry to be so late, Dr. Curtis."

The pert young physician swiveled in her seat. "How is Debbie responding to inositol?" she replied.

He slumped in a chair beside her, too tired for surprise. "You find this case interesting?"

"I do. Dr. Rosenberg has mentioned your research often."

"And how much have you absorbed from Debbie's file?" he quizzed, and folded his crossed hands under each armpit, and effort to silence his stomach growls.

An opportune expression came into her eyes. "I think her condition is treatable, her prognosis is good, and your formula is destined to reap more wonders than anyone has dreamed."

An extended moment of silence lingered as Chip evaluated her remarks. She's good, well versed to handle the pressure, he thought. Finally, he deeply sighed. "How would you respond if I asked you to consult with me on this case?"

A smile spread across her pristine face. "When do we start?"

Chip raised his index finger and pointed to his computer then clicked on a logo. As the window opened, he faced Dr. Curtis.

"Debbie has aggressive hairy-cell leukemia. She is receiving dosages of inositol, and saturation is commencing—I'm concerned as whether she can tolerate added dosage, which may be necessary."

A screen on the computer monitor showed the last blood analysis and the level of inositol in the cells.

Dr. Marcy Curtis evaluated his comments as she reviewed the numbers, her alert eyes scanning the columns. "I'd say the next twenty four hours will unveil the viability of your treatment."

"My prognosis exactly—how about a snack in the cafeteria?" he ventured."

"My treat!" she replied.

***

Two physicians sat in the hospital cafeteria each with a hot cup of coffee nestled in their hands and a cinnamon roll on a saucer. Chip had noticed the beauty of Dr. Curtis for the first time, and she had memorized his handsome face ever since they first met. Yet they were both professionals. Marcy spoke first.

"I understand that the hospital is engaged in securing a company to market your product."

Chip had no response; his mind was elsewhere, mostly at Debbie's bedside as his thoughts drifted from lack of sleep. And finally his mind released his tongue. "Yes, Dr. Rosenberg advised me yesterday, a purchase should be complete by late November, and then there are contractual construction issues."

"We'll, we should complete this statistics study by that time, too." She predicted as she gazed into his eyes reflecting a man seized by his desire to cure cancer and the reality of possible failure. She placed

74

her hand on his wrist. "It seems that you will be a famous man, Chip."

He sipped his coffee. "I'm not the least bit interested in fame. I want my father to receive the honor—since I was a small kid it was his idea."

The long awaited moment of action had come; she would not let it pass. She reached over and touched his hand, and kissed him on the cheek.

Chip was not embarrassed. He needed her friendship, especially now. He raised his head for a moment and looked into Marcy's deep brown eyes, a flood of inhibitions at risk. "Listen, I don't know how much you've heard of Beverly Barrow, but she means nothing to me."

"Shush," she said, with her index finger placed across his chapped lips. "I don't know Beverly, nor do I care to know her—it's you I want to know better."

Chip slumped back in his seat seized by indecision. "I . . . I don't know what to say."

Marcy knew Chip was the humble person she'd searched for since she had left graduate school. No other man would admit he had nothing to say especially after she had fawned over him. Since the day that Dr. Rosenberg had told her that she was to work with this man in his research, she had set her heart to understand this complex Dr. Chip Cook, and now she was satisfied. He was a kind, intelligent, and compassionate man—a man that any sensible woman would love to date, and she was the sensible type.

Marcy smiled at his shyness. "You think we could talk this over one night over dinner?"

His face flushed with embarrassment, yet he really wanted to trust this woman. "Yes, I'd like that very much. And this time it's my treat," he smiled unblushingly.

Chapter 18

A LINCOLN SEDAN drove through the dark evening streets of New York City heading for the bar located across from Cook Medical Center. The driver had had a hard time all day interacting with doctors that gave him little exposure into their offices. He'd never worked with anyone as hardheaded as some of these newly graduated medical doctors, so idealistic, so naïve—no touch with reality. Yet, that was his problem, too—searching for reality. And perhaps he had found the source of happiness in Beverly Barrow.

It was time to relax tonight with Beverly, that is, if she had followed the instructions on the note he'd left pinned to her pillow. The next stoplight cycled "red" and he applied the brakes. The news on the radio was all about Dr. Chip Cook. Besides the news of a clinical study now underway, it seemed that someone had placed a bomb in the famous doctor's car.

The light switched to "green" and he motored through the busy intersection. Three more stoplights shone "green" and he saw the lights of Cook Medical Center up ahead. There were so many lights around the facility it looked like a baseball stadium ready to play-ball. A little obscure bar appeared in the shadows of the giant facility. He parked the car in the off-street parking lot.

He opened the door and punched a code on his key as the horn confirmed the lock command with two quick beeps. The sky was overcast and inclement; a cold front was moving in from Canada and would plunge the temperature into the teens tonight. He tightened his gloves and turned up his collar as he strolled toward the facility.

He approached the entrance door of the quant English-style pub and saw through the large windows an unusually large crowd at the bar and even the booths as he passed advancing to the door. A chil coursed through his body. Must be the cold front, he thought as he twisted the doorknob. He pushed his six-three frame against the door and walked inside, removed his coat, and stuffed his gloves in the pockets. As he hung the coat on a hanger by the door he scanned the bar. Beverly was not there, nor did he see her in the line of wall booths. Assuming that she had not yet arrived he decided to get one of the two remaining barstools. No sooner than he'd sat down, a woman stepped from the restroom. It was Beverly and his face quickly spread into a smile. He stood and waved toward her, who had already glimpsed his tall frame. They met and embraced.

"You came," he whispered in her ear.

She said nothing enjoying the rhapsody cf holding him close. Finally she nodded to a booth now emptying. They slid into the booth; eyes met across the table as he took her hands.

"I wasn't sure you'd be here, tonight."

Her eyes answered his doubts, gleaming with moisture. "You have made me so happy. There's so much I want to ask."

He raised her hands to his lips and planted a kiss on her knuckles. "Ask away."

Her smile was that of a TV commercial about some miracle toothpaste: pearly white teeth, straight, healthy gums, no cavities. She heard his golden voice respond with the first question to set the stage of the conversation.

"I take it that you are no longer involved with Dr. Chip Cook?"

The remark only amazed Beverly at how much the man knew about her past. That's why she was so infatuated with his kindness. Her eyes sparkled. "You know about that, too."

"I know everything about you. Salesmen hear all the gossip—you caught my eye when I stayed over late to deliver some supplies."

"I see. Then you have been spying on me," she grinned.

"Spying: no, attracted: yes. You're a beautiful woman."

The hits were gratifying to Beverly, all the things that Dr. Chip Cook never could, never thought about. Yet she would have given him everything he wanted. But Chip Cook only wanted his research. Oh, if only Chip had seen that, said that; now she loathed the sight of the arrogant bastard.

He gazed into her eyes for a long moment. "Are you hungry?"

"Famished," she responded.

"Let's get out of here. I know a nice place just six blocks from here."

She answered with a beguiling smile and they slid from the booth. He led her to the entry where he found her wrap, and held her coat while she slipped her one slender arm into a sleeve, and then the other. Both hands swished her long blonde hair over the collar. She was radiant, alive, and available.

He donned his overcoat and led her from the bar to his car. Beverly turned before he opened the door.

"Want to walk?" she asked, "The night is young.

Why not," he smiled, stretching the leather gloves over his smallish hands.

They waltzed arm-in-arm along the wide sidewalk almost alone in the coldish night, only two other couples had discovered the quietness of this wonderful place. There was no moon although neither noticed. They approached a long bench at a bus stop and he convinced her to sit for a moment. He put his arm around her and drew her into his chest. His soft, gloved hands swept back her blowing hair. Snowflakes swirled around the bench. They gathered woolen collars around their necks.

He kissed her for a long sustain moment of rhapsody.

All the memories of Chip Cook—really fantasies that never developed—were wiped from her memory. She was totally infatuated, perhaps even in love with this kind, considerate man who had walked into her despondent life out of a majestic fantasy, a hero without a name, only his love for her. Just by his touch a nightmare became ecstasy. Yes, he came out of a dream; he was her hope, her last chance for happiness, to be fulfilled as a woman.

He escorted her into the restaurant and they took a seat at a table in a corner with low light. Beverly sat enjoying her thoughts of a Prince Charming who had just ridden up and kissed her. Prince Charming sat directly in front of her, alive, no pumpkin, no glass slippers. He ordered for them both and told the waiter to bring a bottle of French Wine. As they waited he took her hand from across the table.

His eyes went dark. "You've told me nothing about Dr. Chip Cook than I read in the hospital brochure."

Her dream suddenly vanished, the cutting remark familiar. She averted his face, the hurt, and anger swelling in her breast–she chocked through a sob. "What about that nerd?" she sniffed.

"It's nothing, really–just that I've had trouble reaching him with my products."

She wiped her tears with the back of her hand.

"Yeah, tell me something I don't know–he's too busy with that blasted research project."

"What do you mean by that?"

She deeply sighed, now in full control of her faculties. "He works night and day in the basement of the Research Wing."

He slumped back in his seat. "I see–"

"No you don't see! Cook is only concerned about himself and his work–nobody else–nothing else!" she said contemptuously.

A chill froze the discussion.

Dinner arrived and the hot platters thawed the air. Brian spread his burgundy-colored napkin across his lap with placed his forearm on the table. "I hope you like your steak rare, if not, the waiter can fix that."

She smiled as if lost in her thoughts. "No. Really, I like things in the raw," she replied, and reinforced the remark with a sassy wink.

The conversation drifted into chatter, the steak cut into bite-size pieces, a baked potato mashed in sour cream and butter with a fork. He used lots of hot sauce on the steak and Beverly enjoyed the natural charred taste. He chewed on a large piece and brought the napkin to his mouth. When he finally swallowed, he laid the napkin by his plate and took his fork, glanced up at Beverly as she layered butter on her potato.

"I need to get into Cook's lab and leave some samples–can you help me?" He stabbed a section of potato with his fork, waiting for her reply.

She licked her greasy fingers and wiped them in a napkin. "I can do better than that–I know who has a coded card?"

"Coded Card?"

"Yes, you can access a big safe in his lab where he stores all those fancy chemicals."

He poured two glasses of wine, and Beverly immediately gulped half a glass. He refilled her glass.

"How do I get this card?" he asked sipping his wine.

Beverly cloaked a hiccup with her hand, and drank some water. She sat the water glass down, and gazed into his deep dark eyes.

"Kyle Canfield, his assistant, carries a card in the pocket of his smock."

This dinner was the pinnacle of Beverly's day, a time of relaxation with this newfound friend. A moment of clarity coursed through her mind and prompted a question: was he only a friend or a one night stand? The answer came from the depths of her loneliness: yes, he had rescued her from humiliation in the bar. And yes, he had not taken advantage of her indiscretion. He was the man she'd dreamed of, and Chip Cook would pay for his stupidity.

Chapter 19

THEY MOTORED TOWARD Beverly's townhouse in a sleety snow that swirled in the headlights like attacking moths. Beverly laid her head on Mr. X's shoulder, recovering from countless glasses of gin and soda. The glaring light from each passing utility pole flashed on his face like a strobe light, the lawns now covered with snow. Finally the Lincoln pulled up to the curb in front of Beverly's apartment. He rounded the hood, feet crunching in the snow, and opened the passenger door. He shook her shoulder but failed to rouse Beverly, and he gently slid her limber body from the seat. Her breath reeked of too much alcohol.

He put his arm around her waist, and his glove slipped from his hand and fell to the curb. The annoyed man stretched an arm downward and gripped the glove. He seized the glove cuff in his teeth and inserted a bare hand, spat fibers on the snow. Finally he stood her on the snow-packed grass and wobbly walked her toward the door like a rubber dummy. After they navigated the few steps, he fumbled in her packed purse, found her keys on a ring, and pulled them out. His eyes suddenly focused on a key marked: **Chip's Lab.** Quickly he removed the key and stuffed it in his pocket. He inserted the house key in a brass lock, twisted it, and kicked open the thirty-six inch steel door.

Finally they stood in the warmth of the front room. Beverly's eyes dazedly popped open and she slowly awakened, lost in a hazy dream of blooming daisies in a plush valley. He stood behind her and took off his gloves, one-by-one, stuffed them in his pocket. He rubbed his hands up and down Beverly's torso. Emotions increased, and she undulated in his arms. He kissed her behind each ear, and then moved his hands beside both her temples. He took a deep breath, and gave her head a sudden jerk to one side. The atlas bone snapped, broken nerves disconnected from the brain. Beverly's dead body slumped to the floor.

***

Morning dawned in a blanket of snow as the cosmic sun broke through the spotty patches of snow clouds like a gigantic searchlight. Shafts of penetrating light filtered through the windows of Beverly Barrow's townhouse. Snowbirds sang their familiar tunes perched on leafless branches. Mourning doves hovered together on the icy electrical wires in lifelong pairs. A bushy tail squirrel scratched in a wintry flowerbed searching for pecans he had buried. A hungry cat peered through the sleety-stiff bushes hunting its prey. An owl clenched its claws on a frozen branch, his wandering eyes aligned with a field mouse at the end of its tiny tracks impressed in the snow. Dogs wandered along the fence row sniffing for a garbage can, paw-prints followed the erratic chase.

The cleaning lady, clad in quilted clothing, walked carefully up the steps clinging to the rail as squirrel's scurried up a leafless maple tree. She inserted her passkey into the lock of the steel door. The heavy door swung open as she dragged her cleaning kit into the foyer. She immediately noticed an overturned chair and mumbled as she stooped to erect it upright. And then she spotted a shoe by the chair that prompted more mumbles. A grotesque image brought her hands to her mouth, eyes bulging in a stoic stare.

She screamed!

Unnerved, confused, and terrified, she dropped her cleaning things and dashed out the door, hands waving over her head, wildly screaming. "Police, police Miss Barrow is dead!"

Chapter 20

CARS LINED THE streets of a long row of townhouses just four blocks from Cook Medical Center. Three police cars and a van that belonged to the New York Forensics Lab parked on the curb in front of Beverly Barrow's apartment. Detective Henson was in charge of the investigation and was interviewing the maid who had found the Barrow's body.

Dr. Chip Cook stood in the threshold of the door while pictures were being taken and waited at the foyer entrance. Henson's glance focused on Cook, and he called over a member of his team and instructed him to finish the interview. Henson stepped around the drawing on the floor that outlined the place where the victim had lain. Two forensic technicians were on their knees inspecting something in the carpet.

Dr. Cook stood talking with the coroner when Henson arrived in the foyer. Henson placed his hand on the coroner's shoulder.

"Ben, have you finished here?" Henson reasoned.

"Think so. The woman died of a broken neck. Been dead about twelve hours based on her liver temperature, give or take an hour."

"Okay, I'll see you tomorrow when you've finished the autopsy."

"Yeah," he said and tipped his hat.

Ben grabbed his bag, nodded at Chip, and left through the front door. Chip removed his coat and swung it over an arm. Henson faced him with the same deadpan look when he'd debriefed him on the car bomb.

"This Barrow girl, she was a friend?"

Chip winced, yet not from pain, but his eyes showed some anger as they squinted shut. "You might say that. She worked as the receptionist in the Research Wing. Saw her maybe twice a day."

Henson didn't believe his answer but that was his nature as a detective. Everyone was a suspect until he could tie up the loose ends. "Your fingerprints are on that lamp by the sofa and in the bathroom and kitchen. You want to clarify your statement?" he prodded suggestively.

Chip draped his coat over his forearm. "Beverly allowed me to crash here when I worked around the clock in the lab–just a nice gesture, nothing more."

"You didn't have a misunderstanding with this girl here last night–did you?"

His anger increased. "Now look, I told you what I know."

He acquiesced. "Okay, okay! We'll let it rest there. Incidentally, the forensics boys found a piece of evidence on that bomb in your car but they haven't connected all the dots yet."

He sighed. "That's encouraging but right now I want to know what happened here."

He released his gaze. "Your friend was murdered–that's what happened."

Chip dropped his glance at the chalk marks. "Murdered?" he barked as he swung his coat over his shoulder, his hands numb from the cold. "How can you be sure?"

Henson sighed. "You don't snap your neck by stumbling over a chair."

86

Cook alternately slipped his arms into his coat sleeves feeling more agitated by this numbskull detective than sad, yet not as sad when his when his father was buried. He had never discarded that image in his mind nor discussed the matter with his mother; what was the benefit? It would only bring up old memories, memories that tortured the mind, not remembrance of some joyous fairytale. And now that he'd thought of it again, his life had been a fairytale with haunting shadows of a nightmare. He had lost his father when there were questions that an adolescent really needed answering, answers that only a father could provide. But now he knew why he felt so close to Marcy. She understood without tying him down, no questions about marriage or settling down.

Henson smashed Chip's dream with a slapped him on the back. "Maybe the forensics boys will have some good news."

"Good news? The real question is how many people must die for this cancer formula?" he barked.

Henson gazed into the young physician's eyes for a long moment. He saw depression not rage. This kid was no murderer, he reasoned. He made a mental note to call the forensics lab himself just as soon as he got back to the office. As he released his gaze he inhaled deeply, and clasped the physician's hand.

"I'll keep you posted, Chip."

Chip watched the detective walk out of the door somehow taken aback by his change of attitude. There was too much to accomplish to occupy his thoughts about it now, he had work to be done back at the laboratory.

***

As Chip entered the laboratory, he met Dr. Marcy Curtis at the nurses' station. She handed him a report with her face aglow. He laid the report on the counter and studied the numbers. His eyes suddenly registered the test results.

Many of Debbie's cells were normalizing!

Dr. Curtis took the file and laid her hand on Chip's shoulder. "Congratulations, Dr. Cook. Your formula works." She noticed a tear emerging at the edge of his left eye. "Maybe we should have that dinner tonight."

He managed a smile, but Marcy read his body language. Finally he brushed the tears from the corners of his eyes. "Pick you up at seven, okay?"

"Seven it is," she smiled.

Chapter 21

TWO TECHNICIANS in the forensics lab stared at a specimen through a polarized light Leica microscope, the other at the Nikon camera feeding a Sony monitor. On the desk lay the file on Beverly Barrow, photographs of her body spread on the counter. A sterile plastic specimen sleeve lay open near a pair of stainless forceps. One tech had prepared a slide of a thread taken from the hair of the deceased. The other tech was preparing a comparative slide of another thread taken from the carpet at the crime scene. They had no motive, not even fingerprints; only a small forensic sample, but was it evidence they could use?

An Image-ProPlus software library of natural fibers scrutinized the first thread sample. The tech studied the fiber closely, waiting for the software to ID the sample. The answer flashed to the bottom of the sample: Non-human.

"Well, that eliminates seven billion suspects. What did you get on the carpet fiber?" he asked his partner.

"It's synthetic. We'll have to extract with benzene and run them both by Mass Spec."

He removed the slides and snipped a small piece of each fiber and dropped each in a small extraction bottle with tweezers. From a squeeze bottle he added 700 milliners of benzene. He took the vials to the extraction room and logged them in. Another analyst received the sample and prepared a control sample and a standard sample, and placed them on a shaker, setting the time for thirty minutes.

While they waited for the extractions the two techs studied the photos under a magnifying light. There were no bruises on her neck. Whoever broke the atlas bone knew what he was doing, suggesting he knew something of anatomy–speculation? They discussed the case at length, deciding that they had to identify the threads; the strategy was correct. A buzz from the extraction timer ended the brainstorm.

One technician prepared the extractions for the automation unit while the operation technician punched the codes for the GCMS in automatic run and set the parameters. The samples and a blank along with the controls and standards were all centrifuged, and afterwards were placed in matching 15-ml ampoules using an auto-pipette. The ampoule caps were fitted with neoprene inserts for the auto-injection, and were placed in the carousel married to the auto-injector. The analyzer hummed for about four minutes and an analysis report rolled out of a laser printer.

The analytical tech took the report from the tray, and they both went to the office to see the supervisor. As they walked the hallway, they each puzzled at what the analysis meant.

Brad Rothschild, the supervisor raised his head from a forensics report and rocked back in his chair. "Well, what is your analysis?"

The tech dropped the report on his desk. The summation statement said: The carpet is Monsanto 34-filament nylon, color: Midnight Shadows, approximately five years old. The fiber taken from the deceased hair was still unknown.

"Non-human, back to square one," he replied.

The phone on the supervisor's desk buzzed and he picked up the receiver, carelessly dropping the report on the floor. "Forensics–yes–okay–in twenty."

He hung the phone, smiling. "The criminal unit of the FBI has a fiber specimen they took from the curb in front of the Barrow's townhouse."

"Maybe it will help identify our specimen," a technician surmised as he reached down and picked up the report.

The second tech agreed with his summation. "Yeah, and maybe they know something we don't."

Brad leaned back and pointed his pen. "Don't let these federal boys take this case. This is our town and our case," he barked.

The air went out of the room as the supervisor unloaded his objections. He'd had three cases stolen by the FBI, and all three legal cases were thrown out of court for lack of evidence.

The two techs walked back to their lab discussing the car incident recalling the bomb that the New York City police had removed and had sent to the federal lab. One tech's face wrinkled with something brewing in his mind. Before he could voice his thoughts, a female analyst walked in with her usual smile.

"Hello fellows, having problems I see by your gloom faces. What gives?"

"Oh, hi Prissy, it's this fiber taken from the deceased hair."

They handed her a cover sheet of the report, and she glanced at the analysis, and then looked into their expectant faces.

"Non-human: Have you tried the Reindeer Graphics Fovea software we ordered last month for that environmental case on dead cattle?"

A face brightened. "Yeah, maybe it's still on the hard drive." He raced to the computer, and the second tech followed, but Prissy moved to the coffee pot and looked for breakfast, anything that satisfied her hunger; she always forgot to eat breakfast when on a case.

A technician giddily clicked out of the active program and scanned the hard drive directory. The cursor scrolled down to an icon he recognized, and his index finger double-clicked the mouse.

The file opened.

"There it is, put that slide back under the eyepiece," he advised.

The other tech raced to the counter, retrieved the slide, and ran back to the microscope. He carefully placed it under the polarized light and activated the camera. The camera relayed an image to the microprocessor and the monitor rolled with a series of images one after the other until it decided on a match. An image flashed on the screen: Match, ninety-five percent confidence; the best accuracy possible.

All eyes seized the image: goat hair, specifically *Capra hircus laniger*!

"Barrow was no goat-keeper," one tech replied.

Prissy responded. "Fabric, we're looking for a fabric, you nuts."

Silence captured the room.

The brainpower was immense as the forensic analysts searched their trained minds. Prissy marched back-and-forth, her hand gripping her chin as she studied the fiber. And idea emerged, and she spun around.

"Did anybody find a cashmere glove at the crime scene?"

"No, it couldn't be," they replied in unison.

Brad stood in the door threshold, and walked into the lab. "That's good logic Prissy, it's worth a try. Why don't you hurry downtown and buy us a pair of goatskin gloves."

"Is that all?" Prissy smiled.

"One thing more—the FBI is sending another sample, if it's the same fiber, you may be on to something big. Keep in touch and be careful."

"I will, Brad."

***

Prissy returned to her car, after an afternoon of searching by phone and shoe leather. No commercial outlet in the entire city of New York had in stock a pair of goatskin gloves. She flipped open her cellphone and buzzed the office.

Brad took the call in his office.   "Yes, Prissy?"

"There isn't a pair of goatskin gloves in the city, sir."

"Got any ideas?" Brad asked.

"For starters, the first cashmere goats were imported into the U.S. from Australia and New Zealand about 1990."

"That's a bucket of worms. We need to narrow the search."

"Maybe this will help. Sixty percent of the world's supply of cashmere is produced in China, the rest in Turkey and the Middle East."

"Can't see this is relative, Prissy. With these facts we're either looking for a serial killer or a terrorist."

"Now wait, Brad. I spent an entire day on this project."

"Well, give me something I can use, Prissy."

Her eyes glazed over with moisture. "Cashmere goat farming is completely new in the U.S.—and before you jump to conclusion, I've located a breeder somewhere upstate near Rochester."

"Why didn't you say so—a breeder right here in the state, huh?"

"That's right. He's pioneering another product besides fleece: I think he in the tanning business, too. And I'm off to find this breeder in the morning."

"Good work Prissy. I owe you one. Take whatever you need but be careful, and keep me posted."

Chapter 22

EVENING CHILL SETTLED on the gray streets of the nation's capital city, the sky inclement, automatic streetlights lit in a glowing mist. Winds blew over the rustling waters of the Potomac, gulls wrestled with the high winds seeking places to pitch. Honking horns led the erratic race of bumper-to-bumper traffic whizzing around the expressways. Wrecks piled up at exit ramps going east and west, north and south. The Washing Expressway jammed with commuters exiting the city toward Maryland.

Dr. Robert Caruso, field agent attached directly to the U.S. President, sat in his office overlooking the Potomac finalizing a report due on the President's desk by nine PM tonight. He was late already for a dinner engagement in Rockville related to a terrorist attack in Libya and flipped open his cellphone. As the circuits buzzed he signed the report.

"Thanks for returning my call, Peter . . . yeah, see you tomorrow . . . okay . . . thanks buddy."

Well, the dinner with the CIA was changed to tomorrow, he thought as he sealed the report. A knock on his door lifted his head as a technician came into the room with an unrequested file in his outstretched hand.

"Dr. Caruso, I have a file report that might interest you."

"Can't it wait until tomorrow, Jed?"

"Sure, but the third paragraph mentions a woman by the name Jenny Lynn Cook of New York. I just thought that name would be of interest."

His head jerk up with tired eyes wide open. "Jenny Lynn? Let's have that report, Jed."

He stretched forth his arms and took the report, as his mind rushed back two decades. His hazel eyes scanned down to the third paragraph: *The NY police today removed a bomb from the car of Dr. Chip Cook, director of biomedical research at Cook Medical Center in New York City. The center is named for his father, Dr. John Cook, a prominent physician who died mysteriously twenty years ago. Jenny Lynn Cook, director of Nurse's Training at the center, and wife of the deceased, was interviewed by the police today.*

Robby disbelieved what he'd read as his eyes scanned the column. Then he read a few more paragraphs: *Detective Randolph Henson, NYPD Police Department, gave the results of a forensics test on the bomb in a press conference today, stating that the explosive was C-4.*

He dropped the report musing. His mind replayed a portion of his life before coming to Washington. Jenny Lynn had nursed him back to health, the first time he had met her back in Jacksonville, Florida, where he owned a Chemical Testing laboratory. An accident in his lab had been responsible for taking him to the hospital into her hands. Right away, something attracted him to the lovely creature. They had had a good time together for approximately a year, until he was transferred to Washington by the President to be on the NSA staff. Jenny Lynn came with him on her two-week vacation, but she had disappeared a few months later. And now he had found her again. Fine time, he thought, at the seasoned age of fifty-eight.

His gaze vanished with a smile. He quickly buzzed his intercom. "Jed, come in please."

He stuffed the signed report into a routing envelope as the technician walked into the office.

"See that this envelope is delivered to the President's office by nine PM sharp tonight. And make reservation for me on the next flight to New York–and Jed, I need all you've got on that C-4 in ten minutes."

"Right."

Robby flipped open his cellphone and punched a code as Jed exited the door. "Peter, you've got to stand in for me at the dinner tomorrow . . . . It can't be helped . . . Can you meet me the National Airport in one hour . . . that's great, see you then, Peter, and listen, I won't forget this."

***

Robby made his way through the milling throngs of passengers as crowds of people went to their assigned gates at Reagan National Airport. He stayed close to the side wall as the people coursed around the lobby, while he looked for the café where he often stopped for coffee on these trips.

He finally sat down at a small table and wrapped his hands around a hot cup of coffee. He pulled a file from his briefcase and began reading the contents, which Jed had dumped from the main computers: Two pounds of C-4 in four ounce packets were missing since last year. An investigation had turned up the latest certified withdrawal done by a corporal in the Army, who was discharged six months later; no details on the reason for discharge–the name was overwritten in heavy black ink. Jed wrote a note that someone had sealed a series of records. Robby raise his head from the report and looked for Peter in the crowds.

Peter Meirs, retired CIA assistant to the director had a dark-skinned, gloomy face with jutting jaw and barbed-wire eyebrows, the kind of face that mirrored nothing and rarely displayed a change of expression as if his muscles were glued. Some twenty-five years ago he stood on the parking lot of the White House, screaming at his wife who had gone ahead for the car. A speeding vehicle turned the corner, and ran over and killed her. She had stopped and turned at the sound of his voice. Often, too often, he saw her lifeless body laid prostrate on the pavement that day when he had knelt and held her in his arms rocking her like a baby, tears cascading down his paste-like cheeks.

Peter walked past the café window, and quickly saw a waving arm. He backtracked and entered the shop, strolling to the table where Robby sat.

"Damn, Robby. What's so important?" Peter asked as he sat down.

"Jenny Lynn, I've found her!"

His beady eyes squinted. "What?" he barked disbelievingly.

"She's living now in New York, and she has a son," he replied.

"Well I'll be . . . just imagine—disappeared about twenty-five years ago, wasn't it?'

"Twenty. But I've got to see her, Peter."

Peter's coffee arrived, thoughts still pacing through his mind, thoughts of a friend about to make a mistake. Before he took a swallow he began to chuckle. When the chuckle roared into a rare laugh, he sat the cup on the table.

"What?" Robby snapped.

Peter looked into his hazel eyes still chuckling. "Why you old fart, suppose she doesn't want to see you!"

Robby smiled at his old friend, amused at the humor, but not his reply. "Doesn't matter—I'm going anyway," he hedged staring through the mist rising from his hot coffee cup.

"Well, it just might matter if Jenny Lynn doesn't want to see you."

"I'll cross that bridge later. Listen Peter, I need you to search for a missing batch of C-4—here's all I've got," he said, and slid over the file that bumped into his cup.

Peter steadied the cup; a few drops of coffee hit the table as he thumbed the file. "What gives?"

"Forensics was able to trace the missing C-4 to Fort Dix."

"And you want me to find out who took it?"

"That's right." A loud speaker announced the boarding for the next flight. "It sounds like my flight to LaGuardia is here. Stay in touch, Peter."

Peter watched his longtime friend as he rushed off to the concourse with his briefcase dangling from an arm. Robby was pushing sixty, too old for these kinds of trips. But what worried him mostly was the potential trauma if he saw Jenny Lynn again. Over the years, he had watched Robby push the memory of Jenny Lynn to the back of his mind. Somehow, he had filled her absence with his work. Peter still remembered that dismal day when Robby returned to Jacksonville, Florida, and discovered that Jenny Lynn had left University Hospital without any forwarding address. The pain he saw in Robby's face at that time was the same face he saw today. Deep bruises never healed, they hid and callused, until a moment of great expectancy, then burst and bled. That's why he wished he had gone with Robby. Then, what did it matter. Robby was as stubborn as an old mule. But still, he was Peter's best friend and he wouldn't deny his request. He opened the file and sipped on his coffee, cigarette smoke concealed his face.

***

99

High above the cirrus clouds a Boeing 757 spread its aluminum wings for a flight to LaGuardia Airport. First-class seats were full, empty seats were randomly spotted in other areas. The flight attendant had finished her floorshow as she placed the mike in its cradle after explaining the use of oxygen masks and seat positions.

Robby Caruso pulled down his tray and opened his briefcase. He took out the news-clipping photo of Jenny Lynn: somehow she seemed older than the young image in his mind, and then he was no young chicken himself. He forced his memory back to when Jenny Lynn had left his life. It wasn't at all clear only that she never called or wrote until Kim Marshall explained it all to him. She and Kim had been such good friends and it never occurred to him what had caused Jenny Lynn's sudden departure, even after Kim's explanation. It was a female thing that men rarely ever understood. In reality Robby was fond of Kim, but not romantically and apparently Jenny Lynn couldn't handle that possibility. Peter had reminded him often that he was stupid about women, and he trusted Peter's assessments.

The flight attendant's voice cancelled his thoughts. "Dr. Caruso, would you like your usual coffee?"

He smiled. "Why Melissa, still on these flights, I see—yes thank you."

"Black as I remember?" she smiled.

"Right you are," he grinned, as his eyes turned to the double-pane glass window, a myriad of thoughts racing through his mind: the first day he had met Jenny Lynn when he arrived at the hospital, how cute she was, how attentive, a Georgian girl with those sweet southern comments; yes sir, thank you sir, nice meeting you, dahlin'. Finally, he returned the clipping to his briefcase and continued his gaze at the stars outside the fuselage. Somehow his life flashed by his mind, and one memory stayed focused on the screen of his retina; actually it was a picture of a cat formed by the stars. "Oh yes," he thought aloud, as the waitress cocked an ear smiling: it was Ali, the sweet little kitten that Jenny Lynn had given him when they first met—how long—back before the earth's crust hardened, he aimlessly reasoned.

"Enjoy you coffee, Dr. Caruso."

"Thank you Melissa." he said. The words struck a memory harbored in the cortex of his mind. How many times had he told Jenny Lynn "thank you"—"thank you" for being there, "thank you" for loving me—why did she go, oh why? His first love was far off in New York, but now he had found her at last. Was she as restless as he for companionship? Did she remember those days of such joy? And then the words of the newspaper article silenced his romantic thoughts: she had a son! And he would soon look into her face once more. Was he ready, or would he faint as Peter had warned?

Chapter 23

THE CANCER WARD of Cook Medical Center was alive with nurses and aides registering a sizeable group of applicants for a clinical study featuring a new cancer treatment. The many applicants streaming into the lobby could push the total group above the two thousand target.

Dr. Chip Cook walked to the nurse's station with a clipboard under his arm; his mind triggered one question after another. "How many applicants have registered so far, Julie?"

The head nurse scrolled the monitor screen. "I'd say about two hundred bed patients with no total yet on walk-in patients but the numbers are mounting."

"How many beds are available?"

"The administrator gave us three hundred beds, and I'm not sure that's adequate. And he asked if our space was adequate, too."

"What did you tell him?" he smiled.

"Nothing and he gave us this whole floor."

"You're amazing, Julie. And before I forget, I want you to be sure Debbie Nevins has a bed. And Julie, be sure she gets in the test group, not the control group."

"Already taken care of it, Dr. Cook."

"How could I get along without you, Julie?"

She smiled. "Johns Hopkins sent us five hundred walk-ins and a promise of more," she replied jotting a note on her pad. "Oh, before I forget, Dr. Curtis called this morning—she wants to meet with you this afternoon."

"Good. I'm hoping to use her to establish the numbers for statistical significance. The medical profession is keen on significance and we don't want to strike out there." But his inner thoughts focused on their dinner date tonight.

Chip left the ward and took the elevator to the cancer wing expecting to check in with Debbie. He had authorized another treatment this weekend and reminded himself to talk with his research assistant: he would need his assistance in the dosage prep lab as well as Dr. Curtis' keen mind.

He found Debbie's chart as he continued his uninterrupted stride toward her room. He quietly pushed open the door and found Margaret seated with a magazine in her lap.

"Oh hi, Dr. Cook," Debbie said with a smile.

He returned his bedside smile, nodded at Margaret as he lifted several pages on the clipboard chart. "How are we feeling today, Debbie?"

Debbie rallied her best effort and forced a smile, not her best smile, but all she had in her weakened condition. "Good, Dr. Cook. I feel good."

Chip fingered the plastic lines feeding into her arm. "This glucose will help the weakness," he said, and moved to the monitor that recorded her vital statistics. "You will be moving into the clinical study group tomorrow. I think there are a few you kids like you. Perhaps we can bed you in the same room."

A stronger smile spread over her face. "I was hoping there weren't many kids with leukemia like me."

He stared into her youthful face. "Dear Debbie, I think you will make a great witness to the other kids."

Before Debbie replied, Margaret stood and dropped her magazine in the seat. "She has great confidence in you, Dr. Cook—so do I and Clyde, too."

Chip's head gestured Margaret outside as he turned to Debbie. "The floor nurse will wheel you down to the patient floor in the morning, Debbie.

Margaret leaned on the wall outside the room anticipating Chip's report. "How is she, Dr. Cook—I want the truth."

His hand lowered the clipboard to his side. "The truth is that it's too early yet. I've scheduled her with the test group, not the control group, where she will get constant care. One good sign: the leukemia cells are responding to the treatment, some are even normalized."

Margaret's hands covered her mouth and she began to sob. "Oh thank God!"

Tears rolled down Margaret's cheeks as Chip placed his arm around her shoulders.

Chapter 24

THE MORNING AIR was brisk, the air humid and cold. An overnight light drizzle had melted the snow and left the streets wet. Barking dogs pierced the silence as they chased a stray cat that safely scampered up a telephone pole. A few hungry dogs overturned a garbage can; newspapers blew down the street, hung in bushes and caught in corners. A lone mutt sniffed around a flowerbed and hiked its leg on the grass. A school bus suddenly pulled to the corner stoplight as its screeching brakes startled the dogs, and they dashed into an alley. An early morning jogger crossed to the opposite side of the street, her arm stretched tightly by a tether that choked the pet. The cat sat calmly on the power-line crossbars of the telephone pole and licked its paws purring, as it watched a paperboy toss a plastic wrapped bundle on the front sidewalk in front of a house.

In the misty distance toward downtown, cars lined the street facing a series of townhouses just four blocks from Cook Medical Center, mostly police cars and a Van that belong to the NYPD forensics lab.

Detective Henson was in charge of the investigation and scribbled on a pad as he took a statement from the maid who had found the body of Beverly Barrow. Dr. Chip Cook stood in the doorway while pictures were being taken and nodded at Henson, then waited in the foyer entrance. Henson noticed Cook's arrival, and called over a member of his investigation team, and then left him to finish the interview with the maid. He stepped around the chalk drawing on the floor that depicted the location where the victim had lain. Two forensic technicians were on their knees inspecting something in the carpet.

Dr. Cook was engaged in conversation with the coroner when Detective Henson came into the foyer, and spoke to the coroner.

"Ben, I guess I'll see you tomorrow when the autopsy is complete," Henson said, his hand on the coroner's shoulder.

The coroner bobbed his head and dismissed himself. Dr. Cook removed his coat and slung it over his shoulder. Henson's face had the same deadpan look as when Dr. Cook was in his office debriefing on the car bomb.

"The Barrows girl is a friend of yours, Dr. Cook?" he asked as he draped his coat over a chair rail.

"You might say that. I slept in this townhouse overnight on several occasions in the past month—it's just four blocks from my lab."

"Your fingerprints are on the lamp by the sofa and in the bathroom and in the kitchen."

"Does that bother you?" Chip replied.

"You're saying you slept on that sofa?" he sneered.

"That's right," he barked.

A silent stare froze like an icicle.

"Okay, if you say so, Dr. Cook," he sneered. "Incidentally, the forensics boys found a piece of evidence on that bomb in your car, but they haven't connected all the dots yet."

"Well that's encouraging, but right now I want to know what happened here."

"The girl was murdered."

Chip's forehead wrinkled, a thousand thoughts hid within the creases. How many more people must die for this cancer formula, he thought? First his father and now Beverly—how many more must die . . . how many? Suppose there was only one killer: The Judas?

Henson gazed into the young physician's eyes as he took his coat from the chair. "If you have planned to do something rash, I think it best if you leave the detective work to the professionals."

Chip considered Henson's words as he left the room. Who had killed Beverly was the real question, but the fact that she was dead stuck in his throat. It nagged him so severely that he determined that he just might hire a private detective if Henson had not closed this case very soon.

Chapter 25

A TALL, LEAN vendor entered the front doors of the Cook Medical Center and took an elevator to the radiology imaging area. He was familiar with this wing having serviced several pieces of imaging equipment. The CRT unit down the corridor had a GE processor that he had installed. This afternoon he was prepared to offer an enticing bonus just to learn a few things and gather some information.

He passed several nurses and technicians in the corridor, and one physician. He had his eyes on the pocket of the physician's smock pocket, and purposely bumped into him spilling equipment on the floor, and coffee over the front of his smock.

"Oh, I'm sorry, ah . . . Dr. Canfield," the vender replied gazing at the name badge hanging around his neck. "It was my fault, are you okay?"

"Of course—forget it," he replied and peeled off his smock, lay it over a counter, and rushed into the restroom.

The vendor quickly removed the electronic card in the top pocket of the smock, inserted a fake card, and followed the overhead signs leading to the cafeteria but saw a nurse step into the hallway through an office door. He thought he recognized her face for some reason, and when she spoke he remembered.

"Brian Latham—how nice to see you again!"

He disbelieved his eyes but recognized the voice. "Ms. Cook, what an honor."

His acquaintance with this Training Director of nursing procedures went back to his pre-med days when he had taken a class under her tutorship. She had taught a course in the proper procedure for locating veins in elderly patients, how to insert and inject without damaging tissue, all with the least pain for the patient.

After it became clear that he lacked the brains—mostly attitude—of a medical doctor, his uncle, a respected physician, advised his pursuit in pharmacy. Yet, his uncle withheld his influence, and he dropped out of school during the first year of medical school. Even pharmacy had no interest, and he opted for a position in sales peddling medical imaging equipment and supplies.

Jenny Lynn Cook was a kinder influence in his tormented life, and he was drawn by her kindness, a kindness that had evaded his youth after his mother and father were killed in an automobile accident. Her son, Chip Cook had breezed through medical exams and he was jealous of his accomplishments, especially his family relationship with just a nice mother: his father, Dr. John Cook had signed the final document that had officially terminated his medical training. That event had crushed his ego; he became psychotic, revengeful. It was only because of Jenny Lynn's motherly instincts that he had decided to resign his medical training and depart without recourse. And so, he had joined the Army.

"It seems that you have been successful in your chosen profession, Brian."

He parked the case. "Keeps me outside and traveling, which appeals to me—how is Chip?"

She extended her hand. "He is well, thank you." She paused, a memory flashed in her mind of how

Brian's uncle had acted so suspiciously when Brian tried to reason with him.

"Brian, medicine is not for everyone. I'm glad you have become stable in your choice."

He forced a false smile as the woman he respected waved and went her way. Brian bowed his head, stirring his foot on the floor. Suddenly he slammed his fist on the wall, releasing his angry, yet no one was in the hall that saw his outburst. Careful, don't blow this chance, he thought. He finally decided she was right: medicine was not for everyone. He had planted the seeds of evil revenge, and somehow he felt dirty, filthy, and that only because he had stood in the presence of innocence. Without any means to resist the recurring anger and pain, he had suffered a series of neurosis that had driven him to an attack of neurasthenia. A serial of nightmares kept him awake at night; and yet, as Jenny Lynn had noted he looked at times balanced and normal.

He gathered his strength, and pulled the case toward the cafeteria needing something stronger than coffee but caffeine had to suffice. To his pleasant surprise, Dr. Kyle Canfield sat alone at a table. He quickly filled a cup with coffee and paid at the concession stand, and moved his tray to the table where sat Canfield. He parked his sample case, balancing the tray on one hand.

"May I join you?"

He nodded affirmatively, returning his eyes to his clipboard.

Brian clasped his hands around the warm Styrofoam cup desperately searching for the words to open a conversation with this impersonal stone-man.

"How is Debbie doing today?"

Canfield's head raised, eyes curiously stilled on the question. "Do I know you?"

110

"Of course you must, I just spilled coffee over your clothes."

"Why yes, now I remember."

"I deal in medical image processors chemicals and film."

Canfield failed to shake his hand but responded with a useful reply. "Good. My supply of film is about exhausted."

"That's why I'm here."

His reply was acknowledged with a grin and his eyes returned to the clipboard.

"Those are results of Debbie's blood tests I presume?"

Again he raised his head, this time with a wrinkled forehead. "How is it you know so much about this case?"

"I know Dr. Chip Cook and his mother Jenny Lynn."

"I see. Then you know I can't discuss a patient's records."

He smiled, realizing he had been too intrusive. What did it matter? He had what he wanted, the special cared. "Why don't I leave you a case of free processing chemicals," he replied.

Canfield dropped his fork, and managed a smile. "Thank you that will be nice."

Brian gripped his sales case and left, looking for the elevators.

***

The elevator stopped on the ground floor of the Cook Medical Center. Brian Latham stepped out and walked down the hall to the next corner. He looked both ways, and then flipped open his cellphone and pushed a coded key. As he waited he reviewed the location map posted on the wall. His information said Chip Cook's laboratory was on the basement level with a hallway entrance that had two entrances doors; one from the outside which would be locked, he assumed and the other door into the parking lot of the parking garage. The cellphone buzzed, and a voice spoke with a question.

"Have you got the pass card?"

"Yes. And the office keys, too."

"You've done well. Too bad your cousin failed to enter the safe."

Chapter 26

TWO TECHNICIANS in the Rochester, N.Y. forensics lab stared at a specimen, one through a double-lens Leica microscope, another through a camera monitor with printout provisions. On the desk lay the file of Beverly Barrow with crime scene pictures spread in a line on the cabinet. A plastic sleeve lay open near a pair of stainless forceps while one tech was preparing a slide of a specimen taken from the hair of the deceased. The other tech was preparing a second specimen of the carpet where the body had lain.

A computer program was scrutinizing the carpet fibers of the first sample. The technician studied the fibers through a magnifying lens on the Leica microscope while waiting for the software to identify the specimen. The monitor screen suddenly flashed with a name under the fiber: *Non-human.* The technician smiled. "Well that's a blast: tell me something I don't know," he whispered.

He removed the slide and snipped a piece from the end of the fibers. The fibers fell into an ampoule for GCMS analysis. A squeeze on the pipette of the 1-normal Hydrochloric acid filled the ampoule to a line etched on the glass. He tightened a screw cap fitted for injection by the GCMS auto injector. He punched a code into the keypad and the injection tray rotated to the sample port of the analysis control sample. The injector lowered and pierced the neoprene cap insert. After three seconds, the tray rotated to the ampoule that contained the specimen. Lights cycled, motors hummed, and the laser printer dumped an analysis sheet in the printer tray.

The technician gripped the sheet and read the report: Monsanto 34-filament nylon. Color: Midnight Gloss. Age: 5 years. The report confirmed that the fiber in the carpet was not a match of fibers found in the victim's hair. This could be important evidence, he thought.

The ringing phone interrupted his thoughts. He dropped the analysis sheet and answered the phone. "Forensics–yes . . . uh-uh–in ten? Okay."

She hung the phone gazing at the puzzled face of the other technician.

"What gives?"

She twirled her hair thinking. "The criminal unit of the FBI has a specimen they took from someone's car–"

"Oh, that would be Dr. Chip Cook's car," she snapped.

"They seem to think it's worth a comparison with our samples."

The technician grinned and moved to the GCMS where he had started a sample from the victim's hair. The analysis was complete and he lifted the report from the printer tray curiously staring at the results.

"What do you make of this?" she puzzled.

The female technician took the report. Her examination resulted in similar puzzlement.

"Animal hair: source unknown. That's peculiar. I'd suspected a cat or dog but the analysis doesn't see that–say, what about that software we ordered last month on the environmental case with cattle out west?"

"Yeah, let me see if it's still on the hard drive." He sat down at the main computer and quickly typed several coded messages. The screen opened with the hard drive directory. He scrolled done the list finally recognizing a document. The mouse pointed to the file and double clicked and the file opened. "There it is, put that slide back under the eyepiece."

Gloved fingers positioned the slide. The camera lights lit and relayed an input to the microprocessor: one engine chip routed to another engine chip at lightning speed. The screen on the monitor rolled with sequencing pictures and finally stabilized on a match: Cashmere goat.

Two pair of eyes stared at each other in amazement. The spell was finally broken, and one technician voiced his conclusion.

"That poor girl was no goat-keeper. We're back to square one."

Suddenly a female walked into the office and answered their question.

"Fabric, guys. You're looking for some type of fabric," Prissy Shelly replied matter-of-factly, a pert blond with long hair tied in a ponytail. She had been a technician at this laboratory since she graduated. She had contacts with the FBI forensics group, and the NYPD. She had often analyzed samples needed for a second opinion for court cases, and was the best field technician on the team.

The brainpower was immense. Two forensic technicians were searching their trained minds. As they analyzed their thoughts, Prissy asked a question.

"Did someone loose a Cashmere goatskin glove?" she asked.

Both heads turned. Cashmere!!

Brad Rothschild, the lab supervisor stepped into the room. "You may have something, Prissy. Why don't you hurry downtown and buy a pair of Cashmere goatskin gloves?"

"Is that all?" Prissy smiled smartly.

As Prissy walked out of the lab, she passed a gentleman with a pouch marked, FBI Sample. Her free spirit released her feet in a skipping dance. Now she had a case that required her expertise.

Chapter 27

DEEP IN THE basement of the Cook Medical Center a man dressed in black tights entered the side door connected directly from the parking garage. The time was about 6:00 pm, and Chip Cook was out having dinner with Marcy Curtis. When the coast was finally clear of traffic he inserted a passkey into the lock of the office door, and walked into the laboratory. He strolled to the wall and faced a huge safe. The intruder removed a plastic card from his pocket, and inserted it into a slot on the wall beside the safe door.

A green light lit!

Then he turned the locking wheel.

It moved!

With relieved anxiety, both hands spun the wheel until it stopped, and strained against the massive door as it slowly opened.

Overhead lights came on automatically and he walked into the vault not knowing exactly where to start. He opened trays and searched shelves, and then it occurred to him that the passkey might open a door in the bank of small wall safes. He inserted the key into the lock of three safes without success.

The fourth safe opened!

He pulled out a steel tray and lifted the lid. He found a round metal canister with a handle much like paint cans. He slowly opened the can. There lay a sterile bag labeled, "IP6."

Nervous hand took a glass vial with neoprene-septum from a shelf, and a 1-inch needle syringe from a box next to the vials. Carefully he opened the bag and inserted the syringe needle below the liquid, pulled the plunger to the 10-cc mark. He inserted the needle into the septum and released the contents into the vial. His eyes gleamed as he replaced the bag into the canister. After he locked the tray he reviewed the area, locked the wall safe, and placed the vial in his pocket.

Satisfied that he had a verified sample of Cook's IP6 formula, he quickly closed the safe with a turn of the wheel. Then the intruder dashed to the office door, stepped out into the hallway, and whirled to relock the office door. He reached inside the door and fingered the light switch and suddenly remembered they were left on. He smiled, his mission was achieved.

The black shadow of a man ran down the hall to a set of stairs that led up into the garage parking lot, and reached his car huffing and puffing; he was not physically fit, and perhaps he would use the money paid for this job to join a body-building club, he thought.

Chapter 28

DETECTIVE HENSON SAT in the basement lab of the Cook Medical Research Center conversing with Dr. Chip Cook. He held a file folder in his hand as he leaned back in a chair. "So this is your 'man cave' where you are researching that cure?"

Chip rocked back in his leather chair. "Yes. And we may be closer than ever before to isolating a cure for cancer."

He scoffed. "I warned you that you were dealing with a motive for murder."

"So that's why you phoned me for this meeting, isn't it?"

He uncrossed his legs and tossed the folder on the desk. "I have more evidence but no suspect."

Chip scanned the file. "That's it? You think our killer wears goatskin gloves," he asked and rocked forward in his chair. "As we say in research: This is not statistically significant."

Henson unfolded his crisscrossed arms and opened another file in his briefcase, which he'd received yesterday from the FBI crime lab.

"Maybe not in your line of work but in my work it's solid evidence. Those same fibers were found on the bomb placed in your car, and on the carpet of the victim's apartment."

Chip bounced to his feet, gazed at the data on the file pages, and then closed the folder. "You're saying the person who wore goatskin gloves murdered Beverly Barrow—that's rather thin, Detective."

He bobbed his head. "Don't forget the bomb! The same killer placed it in your car."

Chips eyes gleamed. "How many people do you suspect wear that type of glove?"

"Can't tell yet, a forensics technician over at the Rochester lab is searching the apparel stores in New York at this moment. The FBI is involved now, and a Washington expert of some kind is coming, a Dr. Robert Caruso."

Suddenly Dr. Kyle Canfield rushed into the office. "We've got a problem, Dr. Cook!" he barked.

Henson raised his investigative head. Cook spun around in his chair. "What is it Kyle?"

"Someone has stolen my pass card!" he angrily screamed. This thing is a dummy," he exclaimed and tossed a plastic card on Cook's desk.

Henson swapped stares with Dr. Cook. "Maybe we should open that safe," Henson suggested.

Chip agreed and bounced from his chair. He inserted his card, and opened the safe. Three men entered the safe, but before Chip touched anything Canfield handed him a pair of vinyl powdered gloves.

Henson's face spread into wide smile. "You guys handle this. I've got to make a call," he said, and as he stepped out of the safe he heard the desk phone ring, hunched his shoulder, and picked up the receiver.

"Dr. Cook's lab, Henson speaking," he answered.

"Detective Henson, this is Prissy Shelly. I came up from Rochester to fill out some forms, and wanted to ask a quick question of Dr. Cook."

"Listen Prissy, Dr. Cook could use your forensic skills. Could you grab a bag from the lab, and rush

120

over to Cook Medical Center, we're in Cook's basement lab . . . then I'll answer your question." He smiled musing. His Captain often used the Rochester forensics laboratory when cases were backed up at the NYPD, which was more often than he desired. And Prissy was the best technician in her field.

"Sure thing, be right over."

Henson hung the phone and marched to the safe door. "Hey in there, Prissy is on her way over."

Chip nodded, after he had donned vinyl gloves and unlocked a tray. The tray slid out and he placed it on top of a stack of file boxes. As he opened the lid, he decided not to touch anything until Prissy arrived. Cook and Canfield filed out of the safe and went to the desk.

Henson had poured three cups of coffee from a heated pot next to Cook's desk. They all sat down, and Canfield rolled a chair from the opposite side of the room. As he sat down and took the coffee cup, the steaming liquid reminded him of the accident in the hospital yesterday. Suddenly his glowing eyes indicated he had remembered something, and whirled to the two men.

"I know who took the pass card!"

Henson's eyebrow arched, and Cook stood. "Well speak up, Kyle–who?"

"That vendor . . . Brian Latham–said he knew your mother."

Cook stared into Kyle's face as he picked up the phone and punched a code to his mother's cellphone. While they mulled over the information, the elevator out in the hall opened, and Jenny Lynn walked into the lab.

"I got your buzz as I entered the elevator, Chip. Could you spare another cup of coffee?"

As Kyle poured a cup, Chip escorted his mother to a chair. He looked anxiously into her face. "Who is this Brian Latham?"

She sighed and leaned back in the seat, a million regrets coursed through her mind: why hadn't she told him he was not the biological son of John Cook, why she had not listened to Dorothy Millhouse's advice, and why she had fallen in love again with Robert Caruso. She rolled the coffee cup with her fingers.

"Brian was a first year medical student who was in one of my training classes."

"That's it?" Henson asked.

"That's all there is to it. Brian had too many medical problems, lots of headaches, which took him out of class. He left the medical profession and went into the Army, I think." She knew much more about his problem but it was confidential. Yet nothing in his records would suggest he was harmful, not even his service record.

Chip sat down and rolled his chair facing his mother. "Kyle says Brian stole the pass card to the safe."

Jenny Lynn's face wrinkled, not from the accusation, but his record, and decided to tell what she knew.

"Brian's uncle financed his schooling. I've not looked at any of the financial records but his uncle is Dr. Kosaku who serves on the Board of Directors."

Chip pushed back his chair, his mind focusing on the silent man on the Board. Henson visualized some useful clues with nothing to fit the pieces of the case. But his experience suggested that time was a great puzzle solver, even if these clues were circumstantial—and this case was convoluted twice over.

The hum of the elevator door interrupted the group as Prissy walked into the lab. "Is this some convention or something," she said, suddenly laughing.

Chip and Detective Henson met Prissy and ushered her to the safe. Canfield stopped Chip with a hand on his forearm. "Think I had better get back to work, have a series of dosages to prepare."

Cook agreeably smiled. "Thanks for coming in, Kyle. You've given us a lot to consider, and why don't you take my pass card until we can work out something."

"Sure thing, I'll be certain to return it," he replied, taking the card from Chip's outstretched hand.

Prissy opened the forensics bag and donned a pair of vinyl gloves. She took a black light from the bag and noted that the battery was fully charged. As she coursed the beam over the tray and the sample canister, she gazed into Chip's face.

"The watch-word is cashmere gloves."

Chip's brow wrinkled as he thought of Henson's report.

He wagged his head, hoping that Henson had solved this case before some government agency interfered with the clinical study—perhaps this Dr. Caruso, he thought.

Prissy noticed a smudge, not one but several. Again she reached into the bag and took a roll of special tape. "Whoever opened this case wore no gloves—there're two nice thumb prints here," she announced as she peeled off the prints left impressed on the tape.

Chip glanced at his mother. "What shall we do if these prints belong to Brian Latham?"

Jenny Lynn hung her head. "Let's wait for the results, son."

Henson stood. "Wonder if I could see those records you have."

Jenny Lynn wagged her head. "Not without a court order detective."

Chip Cook remembered her mother's action when his father's relatives wanted to cremate his body. She stood firm until the lawyers agreed only to perform an autopsy which was inconclusive as to cause of death. She had always believed her husband was murdered and wanted no potential evidence destroyed. Chip admired his mother immensely, and now suspected that this Brian Latham had admired her, too.

Chapter 29

PRISSY LEFT COOK Medical Center, the fingerprint samples secured in her bag with a chain-of-custody, and drove back to the laboratory. As she entered the expressway she thought how important these samples were in solving this complex case, and it pleased her that Detective Henson had called her to test these prints. It suddenly occurred to her that Henson was unmarried though she had no reason for such a thought, yet she supposed she did because he was a nice looking guy. She wagged her head and set the cruise control.

It was mid-morning when Prissy arrived at her NYPD Forensic lab. She went straight to the supervisor's office and reported in. And then she took her sample bag into the analysis lab, and prepared a tape of thumbprints on a stereo slide. She placed the slide under the microscope and focused, and then flipped a switch that flashed the image in the microscope on a wide wall screen.

The two fingerprints were obviously identical. These hands did not belong to a laborer, no one who worked with a shovel or an axe, she thought. Quickly she sent the prints through a police databank of prints. The screen was a blur of prints flashing by a scanner which would show the results of the search. The search was nearing an endpoint and Prissy engaged the CIA databank. It seemed that the prints matched no criminal in the police or CIA databanks. She informed her supervisor and left the scanner rechecking the databanks.

***

Prissy left downtown after an afternoon and evening search by phone and shoe leather. She had interesting information but no real evidence. She entered the forensics lab and walked into the supervision's office. She found him sitting behind his desk reading an FBI analysis report.

"Well, here it is but you're not going to like it."

Brad Rothschild gripped both hands behind his neck, rocked back into his chair, and gazed into her blue eyes. "Spare me the dramatics–what have you got, Prissy?"

She sat down and relaxed her tired feet. "For starters, the first Cashmere goats were imported into the US from Australia and New Zealand about 1990."

His arms unfolded. "Well that's a bucket of worms."

She smiled. "Maybe this will help. Sixty percent of the world's supply of Cashmere is produced in China, the rest in Turkey and the Middle East."

"Can't see this helps, Prissy. Either we've got a Chinese serial killer or a terrorist."

"Now wait–I spent all day on this project."

He wagged his head. "Give me something I can use."

"Before you jump to a conclusion, remember that Cashmere goat farming is a relatively new industry in the US."

He smiled with a nod. "Make your case, Prissy."

"I've located a breeder near Rochester who has pioneered another product besides Cashmere fleece."

"And?" he smirked.

She raised an eyebrow. "And, he is tanning goatskin leather."

Brad rocked forward in his chair and stood. "Bingo! Prissy, the FBI sample matches the goat hair sample found on the bomb placed in Dr. Cook's car."

"That's it, then. The wearer of goatskin gloves is our killer. I'm off to look for that breeder in the morning."

Brad pinched his bottom lip with thumb and index finger. "Don't jump to conclusions, Prissy. The fingerprint databank was inconclusive. So take the van or whatever you need, but be careful and report back to me—and you keep your eyes on whomever wears goatskin gloves. This guy is a killer."

She nodded. "Thanks Brad, but I'll be fine."

She needed time off, anyway—the lab had been working overtime with these murders. This was her chance to drive out into the countryside, to be out in the fresh air, and not in that stuffy lab, she thought.

Chapter 30

A TALL MAN with hazel eyes, confident face, and a deep tan lifted his briefcase from the compartment over his aisle seat, and pushed along with the passengers as they exited a Boeing 727 into the lobby at La Gaudier Airport. Dr. Robert Caruso made his way to the rental cars as he checked the time on his wristwatch. It was late morning but he had no appointment. Finally an attendant tossed a key on the counter, and he presented a credit card for payment.

It took only a few moments before he located a blue sedan and he laid his briefcase on the passenger seat, slid into the driver's bucket seat, and inserted the key. It suddenly occurred to him that he'd be smart if he checked the trunk. After he'd inspected the trunk, he strode to the font of the car and raised the hood. His eyes traced the wires from the battery to every electronic item under the hood: nothing. He had satisfied his paranoia, and slid back into the driver's seat. Still an eyebrow arched as he rotated the key. The engine cranked, and he sighed. Maybe Peter was right: he had had too many covert hours under his belt.

It took an hour to reach the main highway, and as he inched along in the noon traffic he reviewed his actions. First he chastised himself for his earlier behavior in the parking lot, and yet, he had faced so many situations in hostile environments where carefulness had saved his life. A smile crossed his leathery face as he mentally apologized to himself. Just why his mind harbored these trivial things that he neither had known nor cared was an unsolvable mystery? All that concerned him now was seeing Jenny Lynn once more—she was the answer to his paranoiac condition, he knew that much.

***

A blue sedan parked in a visitor slot of a parking garage adjacent to the front entrance of the Cook Medical Hospital. Dr. Caruso slid from the seat and reached in for his briefcase. He locked the sedan and his eagle-like eyes ferret out the concrete walls for the elevator. As he dodged around other cars he located the elevators beyond a stairway to the second floor. He passed two medical types, who stood discussing the weather as he reached the elevator. The discussion momentarily tickled his ears as he thought they were reviewing a medical problem—but why was he concerned; indeed, why was he concerned at all? His cellphone suddenly buzzed, and he forgot the question. He cleared his head as he flipped open the plastic nuisance.

"Yeah?"

"Robby, it's Peter."

He smiled. "Did I forget my toothbrush—what's up?"

"That file you wanted on the bomb—it's sealed by a judge no longer on the bench."

"What does that mean, Peter?"

"It means the Joint Chief will have to authorize the file to be opened on the basis of a national security issue, or you can ask the President."

"Not me, not for this—you are former CIA—you open it."

"I thought you'd say something like that, and so I pulled a few strings. The name of the man in the sealed file is a Brian Latham."

"Huh—doesn't ring any bells."

"No chimes here either, but I'll keep looking."

"Who sealed the file?

"Thought you'd ask that, too—a Dr. Kosaku."

Robby nodded, not as affirming the unknown name, only acknowledging Peter's good work. "Thanks again, Peter—I'll be in touch."

"Watch yourself, Robby—you aren't getting any younger, you know."

***

In the visitor's lobby, Caruso strolled to the receptionists' counter and sat his briefcase on the floor. The busy lady behind the counter smiled and raised an index finger, and pointed to the phone receiver in her ear. Caruso nodded and leaned against the shoulder-high counter, while he surveyed the huge lobby. Several people sat in conversation areas, some who waited for their appointments to see a doctor. He saw other salesmen types, seated beside large cases with their heads stuck in a newspaper spread between both arms. He'd seen these types before, who came into his laboratory in Jacksonville, Florida. Somehow he missed those days running the details of a hazardous waste testing laboratory, but deeper in his mind was the question of why Jenny Lynn had left without a word.

Finally the lady asked if she might help him. "Yes, I'm here to see Jenny Lynn Cook, but I don't have an appointment," he said and handed her his business card.

As she read the card: Agent of President Winston Darcy, her eyes bulged wide open, and she quickly picked up a phone. "I have a visitor here to see Jenny Lynn Cook. His name is Dr. Robert Caruso of the President's office . . . yes, isn't that the truth!"

130

She hung the console phone. "Ms. Cook is coming down to meet you, sir, if you'll wait over there," she pointed.

His tanned face smiled. "Thank you."

Caruso picked up his briefcase and marched over to a sofa that faced the elevators in the distance. He sat quizzing his mind about this Dr. Kosaku. Who is he? Why had he sealed the file? Who was he hiding? Suddenly he heard the elevator open.

He stood with his briefcase in his hand.

And then a lovely lady stepped out of the elevator and looked straight at Caruso. She was dressed in a blue smock, and Caruso's mind flashed back twenty years: the night he and this woman had sailed to Galveston, Texas from Jacksonville on his friend's borrowed sloop. He had his first meeting with the President in Houston. What a night it was all alone under the stars in the Gulf.

Jenny Lynn slowly walked toward the man who was the father of her son. Her immediate thoughts centered on Chip, and before this moment had occurred, she had already decided not to reveal that Robby had a child.

The two former lovers met in the center of the lobby face-to-face. It was the moment she had dreamed of for twenty years, the moment she had resisted, and yet his hazel eyes released her inhibitions. And she only realized then, that she had placed her arms around his neck.

Robby felt the warmth of the only woman he had ever loved; the simultaneous shock and ecstasy of holding her in his arms once more. He had searched for her in his mind, in his dreams, in the mirror, and each time a nurse had given him flu short. He would not let her walk out of his life again.

Jenny Lynn grudgingly released her grip and slid from his tall frame until the tips of her toes touched the floor, never losing sight of his tantalizing eyes. These were the deep hazel eyes that had lured her in countless dreams, taunted her as she walked the halls of the hospital, always the subject of conversation with Dorothy Millhouse, when she must talk with someone before she exploded–and the daily reminder of Robby, as she had looked into the same eyes of his son. Reality suddenly struck her as if she'd been slapped in the face. What was her future: Chip or Robby, or eternal loneliness?

Then he kissed her, and she knew it was Robby, the same bliss, the same zing–distant memories flooded anew the deep secrets of her mind. O how she loved this man. Would this moment last for the rest of their years together? Would Robby even consider marriage, or rush off with Peter in another covert episode. Then it occurred to her: neither Robby nor she was young anymore. What a man he was still; broad shoulders, firm in his stance, wavy hair, and those hazel eyes that had haunted her for twenty years. There were a few wrinkles, but he'd stayed in shape, which she recalled was his habit. That's how she remembered him, what she had coveted for so many lonely nights: to be in his arms again–forever.

Her mind finally spilled the thoughts she had harbored for so many years that she had simply lost count. "Oh Robby, we have so much to reminisce. But need I ask you about that affair with Kim Marshal?"

Her words pierced his heart, words that sank deep in the broken pieces of their distant relationship. Was it possible that she thought he was having an affair? "Good heavens, Jenny Lynn. There was never any affair. Kim simply asked that we meet to discuss a case I was working on." he stammered.

Her eyes lowered embarrassingly. Thoughts leaped from memory banks. She had misjudged this poor defenseless man. Kim could wrap him around her little finger. But he had not succumbed to her tricks. Now she knew he still loved her.

"I'm so sorry, Robby. Will you forgive me?"

His manly face blushed. "Forgive you? Jenny Lynn I have always loved. Can you ever imagine what I thought when you left without a word? What's to forgive now that we are together again?"

Tears flooded her face. She wrapped her arms around his broad neck and kissed him, memories surge from her wounded conscience. All these years she had harbored a lie. She held him relentlessly. "How did you ever find me, Robby?"

He stretched back from her arms. The blueness of her eyes stabilized his speech.

"A news clipping came across my desk about a bomb in a Dr. Chip Cook's car."

Her eyes blinked, tears of joy trickled down her cheeks. "I see, so you're still working for the President, huh?"

"It keeps me busy," yet he evaded discussion. "Chip, he's your son, right?"

She, too, evaded the question. After seeing Robby she suddenly lost her confidence, even forgot the advice of Dorothy, but she couldn't mouth those words just now for fear of losing him again. "Is Peter Meirs still in Washington?" said a distractive reply.

He smiled, a memory flashed in his mind of how coy she was when he had first met her in Jacksonville General Hospital, where she was the head emergency nurse. She had acted coquettishly when she undressed him for surgery, yet her soft touch was indelible in his heart.

"Peter is retired; I talked with him just before I left."

She deeply sighed as if the weight of the world had been lifted from her heart. Her fiery blood flowed once again like it did when she was younger, but never like this; was her heart still full of love for this man? O I think so, I know so, she thought. And I also know he must know the truth.

"Well, why don't we go to my office? I have a coffee pot there."

***

Jenny Lynn picked up the phone receiver where she had dropped it when the front desk had called her, and placed it back into its cradle on her desk. She smiled at her only ever lover. He had rushed back into her life, and jolted her existence right-side up. The expression on her face revealed the questions in her mind were perhaps inconsequential. Yet, her unguarded response would be consequential if she took the easy path. She took the obvious one.

"I'll answer your question about Chip now, Robby," she responded with dimpled cheeks. There was a pause, a rush of jumbled thoughts in her crowded brain, how to say the truth without revealing the incriminating biological fact. "Yes, Chip is my son."

"I'd surely like to meet him."

She chucked. "He called me not five minutes before you came. It seems Chip finally has a girlfriend. I thought he would have his head stuck in research for the rest of his life."

Robby grinned. "You know, he's not waiting twenty years before he makes his move."

They both laughed. Joy had returned after a long unscheduled absence. There was hope.

For a long reminiscing time they shared memories about the good days they had had in Jacksonville. The day Robby had brought her an orchid, the trips to the symphony, the days at St. Augustine, when Robby had engaged a visiting artist who sketched a chalk drawing of her youthful face, so young and vibrant, and the hours they had spent walking down the sandy shores of the beach barefoot, each toting their shoes. Finally Jenny Lynn leaned back in her chair and told Robby the entire message of the call.

"By the way, Chip has invited us to dinner tonight to introduce his girlfriend."

"Now I like that—formal or informal?"

"Chip hasn't worn formal wear since he had to dress in a tuxedo when his father was buried—have you ever seen a crushed young man having to wear a tuxedo, when his aching heart wanted to run away in hide in his denim jeans?"

Robby stood and took Jenny Lynn by both hands, pulled her into his arms. "Yes, my father died, too, when I was about fifteen. My mother insisted I wear a tuxedo to the funeral—neither did I like it."

She couldn't resist rewarding him for this tidbit of new information. She irresistibly threw her arms around his neck and kissed him.

Caruso had more memories than he could remember. But those memories now had someone with whom to share. Now he held in his arms the source of all those lost days and nights. The only negative memory was how to keep her from running away again. Jenny Lynn would not settle for living together in lustful existence, she was an old fashion Georgia girl. And that suited him just fine, because he was getting old himself, according to Peter.

Chapter 31

IT WAS A SWANKY location and Jenny Lynn was glad Robby had insisted that he and Chip dress formally. In fact, she was quite wrong about Chip when she'd seen him in a tuxedo; it wasn't the clothes, it was the shock of seeing his father's body. And the girl he'd chosen was quite lovely, too. Marcy Curtis was a charming girl, also a physician; they had much in common, unless the job replaced the home, she mused. She remembered how Chip hated that John was not at home, and that occurrence, too often.

Chip led them to a reserved table in the back room. After they were seated the waiter presented menus. Robby ordered for Jenny Lynn, and remembered how she liked rare steak. Chip relied on Marcy who chose her selection. He glanced down the list and selected Maryland lobster. Marcy chose blackened chicken and a salad. They all ordered wine.

While they waited, the wine was served and they grazed on warm biscuits. Robby broke the silence with a question.

"Chip what's this I hear about a cure for cancer?" he asked as he wiped his mouth with a cloth napkin.

Chip placed a half-eaten biscuit on a saucer and likewise wiped his mouth. "It's not exactly a cure, the product normalizes tumor cells."

"Really? Sounds like a cure to me."

Marcy chimed in with her thoughts. "It's a real breakthrough in cancer treatment. No radiation required," she smiled alluringly into Chip's blushing face.

Chip released his wine glass. "There are still many questions to answer. The authorization of this clinical trial should provide some answers, if not then we'll have to start again."

Robby laid the napkin in his lap and reached for the wine glass. "You know, Chip, I think the world owes you a great deal for championing this study; it seems we've been satisfied with nothing but radiation treatment and new drugs for too long."

"The thanks go to my father," Chip reminded.

Robby gazed across the table into Chip's hazel eyes. Jenny Lynn must be proud, he thought. And it crossed his mind how stupid he had been to let his first love disappear for so long; yet it was out of his control he had rationalized. And he was truly glad that she had Chip to remember. Someday he wanted to hear about his father, but that was Jenny Lynn's choice. He would not push the question, not at the risk of losing her again.

Their conversations were interrupted by steaming plates. As the orders were placed, Jenny Lynn smiled with a thought that had entered her mind.

"What has the President involved you and Peter into this past month, Robby?"

He lifted his arms as a fresh plate of biscuits waltzed by his face.

"Oh, mainly we've been working on immigration. It seems there is a stalemate on completing the fence across the western border."

Jenny Lynn gripped his fork. "What about that plan of yours to open a backdoor to America in California?"

Robby wiped his mouth. "The Senate dropped the ball, and the President was willing to allocate the money from another department, but the press blew the idea out of the water–Chip tell me about this Inositol you've been working with," he said, a ragged attempt of changing the subject. His mouth was sealed by protocol on the details beyond common knowledge.

Chip saw that Caruso was bound to silence and came to his rescue. "It's commonly called Phytate. Why don't you drop by the lab and we'll take a tour."

"I'd like that very much. Thank you."

Marcy and Jenny Lynn knew they had been upstaged.

Chapter 32

IT WAS COLD on the morning that Prissy left her apartment for a day in the country, not a vacation, but a search for a goat farm. On the auto trip she reviewed her notes from a disc she had prepared last night. She plugged the disc in her player and stuck a phone in one ear.

The higher mountaintops caught her gaze through the windshield as she increased the speed of the Mustang up to seventy-five mph. The disc voiced a note she'd recorded last night: "Cashmere goats require minimal shelter and they have their own Cashmere coats for cold weather; it's shed for the summer so they stay cool in the high heat. Not only that, goatskin is hard as nails in wear ability."

The urge of her growling stomach persuaded Prissy to pull in at the next road stop for some breakfast. As the Mustang took the chosen exit and ran down to the cross section, she spotted a restaurant to the right, and pulled into the parking lot. Prissy parked next to a Lincoln sedan, stepped out of the Mustang, and pushed the button on the key. It beeped a familiar warning: her doors were locked.

Only about a dozen people were eating breakfast, and the sausage and bacon smell lured her to the barstools next to a tall, lean, and handsome man. She ordered sausage, eggs, and pancakes because she was famished.   Although she had O.J. at her apartment, she'd forgotten to eat, too excited about her mission.

While she sipped on hot chocolate and waited for the meal, the man's peering eye felt like a fresh tattoo on her neck. She turned. He smiled. As her eyes returned to the steaming cup, she glimpsed his gloves lying by his plate; a red flag flashed in her mind. He'd noticed her stare, and she had to respond.

"Nice gloves."

She suddenly realized her response was incorrect, too casual according to a page in her notes now displayed on the retina of her eyes. His sudden reply caught her off-guard.

"Made for me exclusively, cashmere lined—the best insulator nature has devised."

She ventured a question, and violated the warning that blinked red in her mind. "Isn't goatskin rather rare leather for gloves?"

He released the grip on his coffee cup. "You know goats?"

Prissy stuck a fork in her pancakes suddenly realizing it was her only weapon. "My roommate keeps Boer goats for pets on the roof of our apartment building," she lied.

He chuckled, spilling a few drops of his coffee. "So we're replacing pigeons for goats are we?" he replied, wiping his crotch with a napkin.

They both laughed, clearing the air for meaningful conversation, although Prissy hadn't understood exactly why she had ventured into such a risky situation, except for the prospect of gaining information.

"Prissy Sherry out of Rochester," she said with an extended hand.

His fingers rotated his cup. "I never get that personal," he replied and laid the wet napkin on the table.

"My, this is a chance meeting," she suddenly announced for no reason but to start a conversation of more than four words. For some reason it stuck in her mind how men used one word not twenty, while women used twenty not one.

"How so?" he asked sipping his fresh cup of coffee.

"Well I'm up here looking for a goat farm somewhere in this vicinity—understand they raise Cashmere goats."

He released the cup from his lips. "As a matter of fact, there is a place. It's Billings Farm about ten miles up the road. Why?"

Now Prissy had done it, she frightfully reasoned. She'd parked her butt on this stool beside a possible serial killer, and had talked herself into being his next victim. She quickly excused herself and tossed a twenty dollar bill on her plate.

"Nice chatting with you. I'm running late—keep the change ma'am."

He rotated on the stool, and watched the door as it swung closed behind her exit; his neurotic mind fuzzily replayed the possibilities: Just a coincidence perhaps, he speculated, as he stood and watched a Mustang exit the parking lot. The car stopped momentarily and waited for entry into the line of moving traffic.

He spotted the license plate: NYPD Forensics Lab.

As the man slouched back on the stool his thoughts vanished. His mind was strangled with a tempest of jumbled synapses. He grabbed his forehead between thumb and index finger with a sudden press against the aching muscles. His mind suddenly screamed: Oh why had he forgotten his pills–maybe there was some in his sack, he fuzzily reasoned. He abruptly dashed out the door to the amazement of the cook, and as he approached his car he pushed the key that activated the trunk lid, and it opened like the mouth of King Kong. He stretched open his bag and rattled through a maze of items. There they were: His pills!

He quickly swallowed two of the pills, slammed the lid shut while he stuffed the pill bottle into his pocket. He stumbled back into the restaurant, his mind still throbbing with compulsive thoughts. The ailing man squeezed his neck muscles, blinked his eyes with each throb. He had foolishly prescribed for himself; it was a neurotic mistake. And then he realized in a moment of clarity that he'd fed his persistent migraine with beta-blockers. His reluctance to seek medical attention had reduced his brain cells to a variety of neuropsychiatry disorders. His cerebrospinal fluid often had erupted in seizures at an alarming rate. The pills usually allowed him ten to twelve hours of relief but overuse had lessened its affect.

He finally relaxed into a numb condition as if he floated high above the complexity of reality; no pressure, no pain. His disarranged thoughts somehow steadied temporarily. Even his mind had somehow located a synapse that held the vision of the license plate of . . . of . . . oh yes, Prissy Sherry of Rochester Forensics Lab–forensics, what a fascinating subject he imagined from a deranged perception as a former medical student.

Chapter 33

A MUSTANG PARKED on a sloping driveway at a small office building snuggled against a rising hill. A sign on the front porch announced 'Billings Farm'. Out on the left fenced pasture, a wide, rolling meadow was speckled with grazing Cashmere goats. Prissy swung her purse over her shoulder and almost laughed at her success as she climbed the zigzag steps toward the front door. She gripped the goat-like latch and pushed open the door.

A young lady receptionist smiled as she entered the spacious foyer. "Good morning. How can I help you?" she greeted.

Prissy placed a business card on the counter and the receptionist read the captions.

"I wonder if I could talk to the manager."

The receptionist smiled and bounced from her stool through and open door into a shop of sorts, and yelled. Presently a man of average build, dressed in bib overalls and checkered shirt, stepped through the door.

Prissy extended her hand. "Prissy Shelly is the name."

"Mike Quincy. What's this about, Ms. Sherry," he replied as he released her hand?"

"I understand you manufacture goatskin gloves."

"That's right, so what can I do for you?"

She sized the man, square jaw, tanned face, clean shaven, and he seemed honest enough, a logical conclusion. "I'm looking for someone who habitually wears goatskin gloves, orders them special made, and I just thought you might be able to help."

He placed his hand on his square chin as if he knew something. "Yes, there is one customer who regularly purchases these gloves, perhaps once every six months, I think." He looked at the receptionist for a moment, which caught Prissy's eye.

Perhaps this receptionist had more information than she'd given, Prissy thought as Quincy went back into his shop. She faced the woman, a smallish build, somewhat freckled face, red hair, and a smile the captivated one's attention. "Ah, I wonder if you might have more information—I mean I'm anxious to talk with this man."

She looked into Prissy's face before she answered the question, a moment that seemed to Prissy as illogical, unless she considered some policy that protected the rights of ownership. Finally she spoke.

"Perhaps you had better tell me why you are looking for this man."

Oh, my, Prissy thought, how might she answer that question? Deep within her mind the words finally formed. "Well, my office thinks he might shed some light on a test we are making," she fibbed.

Flags surfaced in the receptionist's mind. "What kind of tests?" she hedged.

Prissy saw something in her eyes, something in the way she'd responded. "It's a cold case—nothing recent," Prissy lied again.

The woman bowed her head and thought for a long moment, her faced a blank. Finally she decided it best if she answered the question. She had harbored doubts about Brian for several years afraid of this very moment, and now it had reached out of the darkness after haunting her consistently. Somehow she knew that questions were bound to arise. Now they had. What had Brian done this time?

"His name is Brian Latham."

Prissy already knew that name, and now she was greatly surprised at just why this receptionist knew it, too, so secretly, yet somehow personally, more so than just a customer. And then it occurred to her: Suppose Latham had followed her? It was a valid question, perhaps with no answer.

"Could this Latham fellow be coming for more gloves anytime soon?"

The receptionist saw the familiar blush of fear on Prissy's face, and knew Brian was in trouble again. She reached out and took the hand of the woman, an impulse she couldn't resist–perhaps she always had hoped that someone cared besides herself. She had never told anyone but her husband, and now concluded that she had better reveal her secret to this woman. The cards were on the table, and she held the joker card.

"Brian is my brother," she confided.

Prissy frowned as her mind failed to compute what the words actually meant, and then she realized it wasn't the words, but the misery she saw in the woman's eyes.

"Oh honey, I'm so sorry."

"The name is Alice Quincy, and I'm sorry, too. Brian has always been sort of weird, especially after our mother died, and then father, too."

"Alice, I must tell you that forensic evidence makes Brian a suspect in the murder of a woman he was dating."

Her hands immediately clasped over her mouth as she wiggled her head, both eyes shut. She had always worried that something drastic like this would be

145

Brian's epitaph. Thoughts of their childhood flashed through her mind like a laser had cut through her heart, and she suddenly understood that Brian had to pay for his mistakes; she had no other answer, the law was clear, her misery confirmed. She had kept him safe for ten years, and now she had no more strength, no more patience, no more hope that he would ever recover from this madness. One thing was sure; she could not pay for his crimes, she had paid enough, never married because she harbored a fear of birthing children plagued with Brian's disease, until she had met Mike. She wasn't even sure it was a disease, but probably the way her father had raised him she rationalized. Either way, it was all over now. She inhaled and deeply sighed. Somehow the endless pressure was relieved. A mountain of worry had been released.

"Prissy, Brian is due here any moment. I don't think it's wise to hang around here."

Prissy took Alice's hand. "What are you going to do now? Is there anything I can do?"

"Well, I just don't know what to do."

"You and your husband could both move in with me—I have nobody, and we can work this thing out together?"

Prissy's caring spirit and the love she had shown spoke to Alice's heart. "I don't really know—its Mike's decision. Please, you must go, Prissy."

Even as she spoke Prissy heard the sound of an approaching car in the near distance. She turned, waved at Alice. "We'll continue this conversation later, honey."

Alice sensed her honesty and wondered if she should have agreed to leave.

Prissy bounded off the porch and opened the door of her Mustang. Quickly she cranked the 289cc engine and scratched off toward the pasture. She followed a dirt road that ran parallel to the black top road, as she frantically looked for a turn that led her to the main highway. When she suddenly stopped, and got her bearings, she saw in her rearview mirror that the Lincoln sedan was in chase. A road sign in the distance pointed left, and she took the turn without a clue of where it went. Her scrambled nerves only told her to hide somewhere because the Lincoln was faster than her rebuilt Mustang.

After she had gone a quarter mile she saw an old farmhouse in the field about a hundred yards away sheltered by large trees and tall hedges. Apparently the house was empty, since the weeds had overgrown the porch. A snap decision forced her vehicle into the driveway, and she drove around to the back of the house. The driveway was loose gravel and left faint tracks. Perhaps, she thought, she might be safe here. She grabbed her cellphone on the seat and rush up four steps onto a creaky wooden porch. The door was locked and she painfully crashed her shoulder against the door. A rotten jam broke apart into splinters, and the door squeaked open. Quickly she rushed through the threshold, shut the door, and wedged a wooden chair under the glass doorknob. She had seen this procedure in movies, gratified that she even knew what to do. Now she waited, her nerves allowed no rest.

Finally she fumbled in her pantsuit and found the cellphone, chided herself for not phoning sooner. Her shaky index finger punched a preset code. The circuits clicked and bussed, and a voice spoke.

"Forensics Laboratory, how may I direct your call?"

"Melissa, this is Prissy–is Brad there?"

The receptionist clutched her headphone. "Prissy! We've been worried sick about you–are you all right?"

"Yes, but where is Brad?"

"Oh! Brad took the helicopter toward Rochester hoping to find you. I think he is looking for a goatskin tanning facility."

"Can you contact Brad and tell him I'm holdup in an old farmhouse about a half mile east of the tanning facility. Tell him it's urgent—Brian Latham is the goatskin gloves killer, and he's following me."

"Great Scott! Right away Prissy, please be careful."

Prissy flipped closed her cellphone and carelessly slumped in an old rocking chair, then arose, and apprehensively walked to a window on the east side of the house. It was boarded up with wide cracks that gave her a view of the road and the driveway. Suddenly she saw a dusty rooster tail rising in the distance behind a vehicle obscured by tall hedges. Was it Brian? Of course it was!

Prissy crouched in a corner near a closet door. Suddenly she heard footsteps on the porch. The door shook, the chair finally fell away, and the door swung open, a haunting squeak pierced the creepy silence. A sinister shadow stood silhouetted in the doorway like a ghost from hell. Prissy unconsciously screamed. The shadow ran forward.

"Prissy, it's me, Alice."

A terrified forensic technician deeply sighed with relief, her heart pumping erratically. "Oh Alice, you shouldn't have come."

She nodded her head, knowing she must tell Prissy what was in her heart. "Brian is my brother, Prissy; maybe I can talk him out of this madness."

Prissy was astonished. Yet she knew from experience that it was difficult at best to predict the actions of a neurotic person, especially one who had committed murder. "Honey I think you had better go back to the shop."

Alice stood helpless, a part of her heart wanted desperately to help Prissy; another part felt the need to somehow save her brother. She had to act, but how?

Suddenly the door burst open and a larger shadow stood on the threshold, its eyes scanning the unpainted walls of his boyhood house. His mind flashed to his youth, bitter memories scattered in the dark corners of his tortured mind. He was born in this house and raised on this farm with his mother and father and one sister. Several uncles had owned the farms in this valley, and a host of first-cousins had played endless hours on the grassy range, including himself.

His eyes blurred as he focused down at two women huddled in a corner. Brian's deranged mental condition clouded his reasoning powers as he starkly recognized both women. Prissy Sherry, the lady he'd met in the restaurant, and Alice Quincy . . . his sister? What was she doing here, his battered subconscious asked, and what was *he* doing here, himself? There was no answer to either question in his tormented mind. He stood on the edge of insanity tittering back-and-forth, lost in a maze of question marks. Childhood pictures flashed across the retina of his irritated eyes, one-by-one, a myriad of distant torments he had long ago forgotten, but the lay stored in his mind. His knuckle's rubbed in the dark sockets of his irritated eyes, a fuzzy image suddenly focused in his mind: Beverly Barrow! A girl he loved and had . . . had killed!

His mind suddenly imploded.

Pain—excruciating pain, gnawed on the meninx protecting the brain and spinal cord. And then just as suddenly, another image appeared in the recesses of his sketchy mind: Jenny Lynn Cook! The memory of this kind woman relaxed the pain in sudden comfort. He fell to his knees; his opened palms screened his eyes from reality.

Alice rushed forward. "Oh Brian, you need help—let me help you, please."

He raised his head and saw not his sister, but his uncle, his devious uncle, the albatross that hung around his neck for tormenting years. His eyes waxed red, revenge seized his helpless mind. Disabled of rational thought, his rage balled a fist, and struck his "uncle" in the head. Instead, he had perilously thrust his sister across the floor, and helplessly thudded against the wall. He stood and screamed like a howling coyote, his hands rose like a zombie, his face a shadow of ghostly horror.

Chapter 34

JENNY LYNN CLOSED her cellphone, fear in her eyes, urgency in her thoughts. Dr. Caruso sat at a table in her office with his laptop open, and noticed the paleness of her face. He'd seen that look before and thought something serious had gone wrong.

"Is there some way I can help?"

She clasped both hands over her face, tears streaming down her cheeks. Robby stood and walked the few steps to her desk. He impatiently waited until she'd regained her composure. She stood and fell into his arms, the hiding place she had learned long ago to trust.

"Oh Robby, so many things are falling apart—it's a miracle that you are here. Your services are needed desperately."

He stroked her blonde hair and pressed her face against his tanned leathery cheek. "What's going on here?"

"Right now there is a young forensics technician who is in deep trouble. Detective Henson just informed me that her last message said a deranged man is chasing her, the same man he believes is the killer of a receptionist at the Medical Center."

"I need some particulars: where is she?"

"Detective Henson is expecting you—he received word you were coming to New York—it's a federal case now," she replied, and tore a sheet from a pad on her desk. "Here's the address."

Caruso took the paper and looked into her wet, but still pristine face, and wiped a tear from the corner of her blue eyes. "You wait for me, we'll have dinner tonight."

She nodded. "Be careful, darling. You're not so young anymore."

He flashed a toothy grin, probably the first time he'd felt this giddy in years. "Don't you hate it? That's also what Peter reminded me—you two are alike in some ways."

"Maybe, but he drinks more coffee than I," she smiled alluringly; his playful remarks in the midst of calamity gave her sweet confidence. And then she remembered that he always had been resourceful in calamities.

***

A rental car drove into the parking lot of the NYPD Midtown Precinct on West 54th Street, located about a mile from the Medical Center. Dr. Caruso rushed to the entrance as he closed his cellphone. Detective Henson was on the rooftop of the building awaiting a helicopter. Caruso lost no time when he saw the crowd of people at the elevator, and took the stairway. Fortunately, the roof was only two floors as a sergeant had correctly instructed.

Dr. Caruso burst open the roof door and suddenly heard the roar of helicopter rotors. Henson, or somebody, waved and motioned a hand. Caruso stooped beneath the rotors and stepped into the cabin. As he closed the door the pilot handed him a headset.

A hand reached from the backseat. "Dr. Caruso—glad you came," said Detective Henson.

Robby nodded, twisting in his seat, and took the handshake of a hand thrust between the seats. "Want to fill me in, Henson?"

He smiled. "The man in chase of Prissy Shelly is Brian Latham."

The name instantly flashed from Peter's conversation. "Sherry is the forensic technician?"

"Yeah, a smart kid I'm told."

Robby's mind returned to Peter's conversation. "I can give you a few facts on Brian Latham."

"Yeah?" said a surprised detective.

"I have tangible evidence that Latham stole a batch of C-4 while in the Army."

Henson slumped back in his seat, pieces were finally coming together. "Forensics has a hair sample from Beverly Barrow's blouse they suspect will match a hair they found in the carpet. If you can pin the C-4 on Latham, and the hair indeed belongs to Latham, we've got a case, but the supervisor thinks the hair is from some goat," Henson speculated.

"A goat? Well that *is* interesting. This Beverly Barrow, is she the receptionist from the Medical Center?" Caruso surmised.

"Yeah, and according to the forensics supervisor, Prissy has evidence that seals this case tight as a drum."

Caruso glanced at the time on his Rolex wristwatch given to him by the President. "I'm interested in this Latham fellow—got a cold federal case to solve myself."

Chapter 35

A POLICE HELICOPTER coursed over the rural, green hills of New York State some twenty miles south of Rochester, a port city northwest of New York City on Lake Ontario. One of the passengers trained his binoculars on the hillsides and valleys. Up ahead he focused on the tiny images of goats chewing grass, their goatees dragging over the clean areas trimmed by their sharp teeth.

The passenger lowered the binoculars and pointed to a landing area in the near distance. As they neared the landing spot, they heard gunshots. Henson trained the binoculars on a Mustang parked behind tall bushes, a rusty truck, and a black Lincoln sedan parked near the back porch, driver's door left open on all vehicles except the Mustang. The whirlybird lowered altitude and settled on a patch of grass twenty yards from the Mustang. And then Robby noticed over beside the barn where a parked helicopter sat unattended. "Something screwy is going on down there," he whispered.

Caruso leaped from the front passenger seat, and reached the porch first, trailed by Henson. Caruso lowered his shoulder and crashed through the closed door. On the floor lay a man in a pool of blood. Henson dropped beside him and placed two fingers on the carotid artery. He had a weak pulse.

Caruso bounded up the creaky steps, when he heard overhead footsteps. As he neared the top newel post, he quickly located the source of the noise in the left bedroom. He rounded the corner, and the door of an adjacent room suddenly burst open. A screaming man rushed out and plunged his head into Caruso's stomach. Both bodies crashed against the banisters. Caruso slammed his open palms on both sides of the man's head. The raging man fell to his knees, his hands pressed over his ears, and then just as quickly he rose on one knee ready to counterattack. Caruso kicked him under the chin. The man fell unconscious on the floor.

Caruso whirled, both hands on the stair rail, and looked down at Henson checking the bloody body on the floor. "Is he alive?"

"Just barely—he's Brad Rothschild, supervisor of the Rochester Forensic lab. Damn fool, he couldn't fight his way out of a paper bag. We've got to airlift him to a hospital."

"Right, after I take a look for Prissy," Caruso replied.

He opened the door on the left, where he had heard the noise. As the door creaked open he saw a young woman on the floor gagged and bound with her hands secured behind her back. Caruso rushed over and lifted her torso against his stooped knee. He untied the gag and the young lass spoke feebly.

"Don't know who you are, but I'm grateful," she gasped.

He stood Prissy on her feet, and released the bounds on her hands. She had deep lacerations on her wrists. "The name's Dr. Caruso, Prissy."

She was too weak to walk, and Caruso cradled her in his strong arms, and navigated the stairs to the first floor. When he stood her on her feet, wobbly hands suddenly gripped over her mouth. She stooped beside her supervisor on the floor. Was she the cause of his death, her brain screamed?

"Is he alive?"

Henson nodded affirmatively.

She sobbed, and then her mind recalled that Alice had entered the house. "Oh, Dr. Caruso–that woman over there is Brian Latham's sister," she said, and crawled over to a still body.

Caruso and Henson followed Prissy. "What's her name?" Henson inquired.

Prissy said nothing as she gently caressed the hair of the woman's head in her lap. Tears rolled down Prissy's face. Robby stooped on one knee and pulled Prissy into his arms. She looked into the face of the man who saved her life.

"Her name is Alice Quincy, Dr. Caruso."

"You can call me Robby, Prissy," he said with a slight grin.

Before he had finished his sentence, a man suddenly burst into the room, his heard turning left then right. He saw Prissy in the arms of a strange man, and rushed over. And then he saw his wife's body on the floor. He slumped on his knees, his heart pounding erratically as he madly screamed.

"Oh no–No!" His cries pierced the morbid silence, like a wounded coyote.

Prissy faced Henson. "It's Mike Quincy, her husband," she sniveled.

Henson placed his hand on the sobbing man's shoulder. "Better come with us, Mr. Quincy. We need your statement," he said, and turned to Caruso. "Okay. Let's get these people to the helicopter. I'll call backup to impound that Cadillac, and they'll send a forensics team out here."

Caruso tied Brian's hands behind his back with Alice's shawl, and led him out to the helicopter. He looped his leather belt round a wheel strut and secured it around Brian's waist. Prissy was assigned to watch the prisoner.

Caruso constructed a makeshift stretcher from the broken boards on the porch. And while Henson made his call, Caruso and Prissy carefully placed Brad on the makeshift lift, and carried him to the helicopter. When they approached the copter, they noticed that Henson's pilot was over by the barn where they had seen the second helicopter, which they now realized was Brad's airlift.

The pilot waved his hands and called out. "Over here!"

Caruso and Henson rushed over to the other helicopter, while Prissy stayed with Brad and Mike Quincy. When they arrived, they saw the reason for the pilot's call: Brad's pilot had been shot dead; another victim of Latham?

They placed the pilot's body in the back seat of Brad's helicopter, and the trio went back to check their patient. When Henson closed his cellphone, he reported the news.

Henson shook his head. "They can't get a team out here for another three hours." He realized the bodies were piling up and wondered if Latham had actually killed them all. Yet it wasn't plausible in his mind; serial killers were methodical. This guy killed when he had a pain in the head, he thought, as he held a bottle of odd-looking pills taken from Latham's pocket.

After Henson had gathered his thoughts on the situation at hand, it was obvious that both copters were needed for transport. Brad didn't have three hours to live without treatment. He voiced a pertinent question.

"The second helicopter, whose going to pilot it?"

Caruso answered the question. "I will."

The pilot smiled. "You're checked out on these things?"

"Taught by the best," Caruso replied. "Would you do me the honor of describing the controls–the dials are in all the wrong places on these relics?"

Henson bobbed his head in disgust, or perhaps stark fear; should he laugh or cry, he thought. Caruso was a man of many talents, and he thought it best to let things develop.

Prissy had noticed that Brad was breathing shallowly, his face pale as a ghost. She checked his pulse and decided he had lost too much blood. He needed a transfusion immediately or he'd die before they reached the hospital. But could they perform a transfusion in the field? Medics of the Army had done it, and now they had no choice. Driven by resolve, she fished in Brad's wallet and found his blood donor card, standard practice for all forensic teams. Likewise she noticed that she had the same blood type as Brad; what a break, she reasoned.

When the men returned she explained the situation, and after a moment of discussion they finally agreed to the transfusion. They surveyed the experience among the group, Prissy had experience in lab procedure, and Caruso had two PhD degrees in chemistry and biology. The medical team had been chosen by default; it was Brad's only chance, yet he had no vote.

"Okay, Prissy," Caruso grinned. "I'm B-type".

"Me, too," Henson added.

The pilot grinned. "I'm out, too–A-type."

"Oh you ole fuddy-duddies, I'm Brad's type," she announced. "You guys find the necessary supplies."

A host of relieving sighs accented the light breeze coursing across the valley. This group was no MASH outfit, but the field conditions were similar. Prissy knew what must be done, and hoped that Dr. Caruso was as good an improviser as a chemist.

The pilot found a length of plastic tubing in his tool case, actually extra fuel lines. Prissy's travel case was a treasure-trove of useful items. She found binder clips, several hypodermic needles, sterile pads, and a small roll of duct tape. She collected the items and laid them on a scarf she found in the back seat of the helicopter. Henson and the pilot fashioned a pallet from a few boards taken from the porch of the house, and laid them beside Brad.

Caruso removed the plunger from each of two syringes and affixed a length of the plastic fuel line onto the two chambers with a twist of wire that he found in the barn. Henson found a half-empty bottle of Scotch, probably left by the farmer. The taste was flat but the alcohol was still antiseptic. Caruso sterilized the connections and the needles with the Scotch and handed one of the needle-ends to Henson, while he fashioned an IV with the other needle. When Caruso nodded, Henson gave his transfer line to Prissy, who lay reclined on the pallet aligned with Brad. He wiped the flesh of her wrist with the Scotch.

Her eyes were glassy, her heart pumping wildly. She took a deep breath and injected the needle into her own wrist. Henson secured the needle around her wrist with duct tape. Deep dark rich blood coursed through the needle and though the line. When the blood dripped from the other needle held in Caruso's hand, he bobbed his head. Henson pinched the line from Prissy's arm with a binder clip. Caruso inserted the needle into Brad's wrist, and Henson released the clip at Caruso's nod. Caruso poured Scotch over the injection point and secured the needle with duct tape. Life-giving blood flowed from Prissy's wrist like water through a burette stopcock. Caruso watched the second hand of his Rolex and estimated the drip flow. He calculated the total time for transfusion at forty-five minutes as he set the drip rate by the pressure he applied to a binder clip over the line to Brad's wrist. Henson released his breath, placed his coat over Prissy, and spread Caruso's jacket over Brad, which he hoped would keep them warm.

While they waited Henson punched a code on his cellphone to his office. Caruso fixed the binder clip at a calculated rate, and briefly walked over to the second helicopter where Brian was seated handcuffed in the back seat. His traumatized body seemed to be in some kind of trance. And so he flipped open his cellphone and called Jenny Lynn. As he waited, he fingered his belt on the floor.

***

At forty minutes into the transfusion, Caruso clamped both ends of the tube and extracted the needle from Prissy's arm. And then he removed Brad's syringe needle, and flushed both injection points with Scotch. He sealed the wounds with a piece of sample-bag tape he had stripped from two sterile bags he later found in Prissy's travel case.

Henson removed the helicopter passenger seat to accommodate Brad's stretcher, while his pilot gave Caruso a brief flying lesson in the cabin of the other whirlybird. He allowed Mike Quincy the courtesy of staying with his deceased wife. Prissy gave Quincy her car keys; he agreed to drive it to the Rochester forensics lab with Alice's body. Prissy road with Brad in the backseat just in case he roused. Robby and Henson were in the second copter with the dead pilot and Latham.

Caruso donned the flight helmet, and hit the rotor switch. He watched the pressure gages and gyro speeds until the rotor speed was nominal. And then he faced Henson in the copilot seat.

"Ready for liftoff?"

Henson said nothing, his hands gripped tightly around a handhold, knuckles white, heart thumping.

Caruso pulled back on the stick. The fuselage lifted, and banked toward New York City. After the helicopter was fully airborne, the pilot in the other helicopter settled in behind Caruso. The two pilots were in contact through the flight helmets, which eased Henson's nerves, but not his stomach; he fought the urge of regurgitation.

The experienced pilot made radio contact with the heliport at Cook Medical Center. After they successfully landed, orderlies removed Brad, and Prissy went with them. Caruso sat down the other helicopter on the Police Heliport atop the Midtown Precinct, where Henson's team was waiting. Henson unlocked the handcuffs from the seat frame, and escorted Brian Latham toward the elevator leading to the precinct jail.

Henson divulged the harrowing experience with his supervisor. He admitted that of all his police experience, these were the most practical few hours he'd ever witnessed. Dr. Robert Caruso was a real hero—was there nothing he wouldn't do, or anything he couldn't do, he thought?

Chapter 36

SEVERAL OFFICERS FROM the Midtown Precinct on West 54th Street gathered around a table in the break room, each took his turn discussing their version of the news. They all had some part in searching for a serial killer, who had been moved downtown. Reporters in the outer room were busy punching cellphones. Information said the police were holding Brian Latham, a psychotic who had killed at least three victims; no word on the names of the victims, yet. An eyewitness, a farmhand, testified that he had seen Latham when he shot Crosby, the helicopter pilot for Captain Brad Rothschild, supervisor of the Rochester Forensics lab. Rothschild now lay in critical condition at Cook Medical Center.

Detective Henson and Caruso had brought the suspect in last night. At a press conference, Henson fielded questions. The press corps sat stunned as Henson relayed the information. A mountain of clues–more than any case they had ever reported–placed the alleged killer in several incriminating locations. Forensic evidence located him in Barbara Barrow's apartment where she was found dead, and a cashmere glove hair was in the laboratory office of Dr. John Cook, also found dead, including the same type of hair on the bomb placed in Chip Cook's car. He was captured just before he had the opportunity to murder Prissy Shelly, a forensic technician in the Rochester forensics lab. He had also accidently killed his own sister, Alice Quincy, and critically wounded Brad Rothschild. A judge had ruled no bail, and assigned a public defender, since no relative came forward, nor acquaintance. His employer had issued a statement: no further comment.

Henson sat at a small table in the interview room of the NYPD downtown police station, and faced Brian Latham; his appointed lawyer sat beside him, cursorily reading papers for his defense. His Miranda Rights were read in the lawyer's presence. Latham grunted, indicating that he heard and acknowledged his rights. His lawyer witnessed that his client had been properly processed for arraignment, but still his client had not signed any papers.

Henson leaned back in his chair and crossed his legs. "Brian, the court will go easier on you if you sign a confession."

The handcuffed prisoner sat stoic, and said nothing as Henson glanced at his lawyer's face. Finally the lawyer questioned Latham, but still he said nothing. Henson's patience wore thin.

Caruso and Prissy stood behind the one-way mirror/window, and watched the procedures. Suddenly Caruso's cellphone buzzed. Prissy's head turned with alarming fear in her eyes painfully sensing that it was bad news about Brad. Caruso flipped open the phone.

"Yes."

"Robby, it's me Jenny Lynn," said a tiny voice located across the city.

Caruso nodded his head and winked at Prissy. "Hi darling—everything okay on your end?

"Well frankly, no."

"I'm sorry, what kind I do?"

"For starters, you can come to my house and take me to that police station."

"Sure, but why?"

"Brian Latham was one of my students."

Caruso's left eyebrow arched as he thought she just might be able to persuade Brian to sign a confession. "Be right there in 30 minutes—make that 15. I'll fly the police helicopter."

"Who taught you to fly a helicopter?"

"Peter."

She smirked. "I can believe that."

***

Jenny Lynn stood on the front yard and held her hair out of her eyes as a helicopter lowered to the residence road in a swanky neighborhood. The skis touched the pavement and the passenger door swung opened. Jenny Lynn stooped and rushed under the rotors, her skirt flapping. She gripped a handhold and jumped into the cabin, quickly closing the door, and brushed her hair out of her face.

Caruso smiled. "Buckle your seatbelt, darling."

She returned his smile. "You bet I will—I didn't know Peter could fly one of these things, and I'm not sure you can either."

He dismissed her sarcasm. "So, Brian was a student—how so?" he yelled above the rotor noise, as he pulled back the stick with his feet on the elevator paddles, and banked left.

She deeply sighed, a memory she had forgotten until she had briefly talked with Brian earlier in the hospital. She clasped her hands together, placed them under her chin, shadowy thoughts forming in her mind.

"It was many years ago. Brian was a medical student and he took a required class on proper needle injection that I taught for the center."

He chuckled. "Maybe I should have taken that course. You should have seen the direct blood transfusion Prissy and I performed on Brad."

164

She touched his shoulder. "I heard. That was a super emergency procedure you guys performed. I'm proud of you, Robby."

He winked. "So Prissy told you?" he replied, as the helicopter maneuver toward the top of the NYPD police station.

"No, it was Henson. You made quite an impression on that detective. I think he wants to invite you out to dinner tonight."

"Well, I'm game for food."

"How I remember," she smiled.

"Okay, hold on to your teeth, we're coming in."

***

Henson opened the interview door and stepped into the hallway. He walked to the entry of the view room, eager to speak with Jenny Lynn. As he entered the small room, Prissy hastily turned the corner and took his arm, her face frantic.

"Brad has had a relapse of some kind. I've must go to him."

"Right," he said, and flipped open his cellphone, said a few words, and then closed the plastic nuisance. "One of the pilots downstairs will fly you over in the helicopter that Caruso just landed," he replied.

Prissy's eyes locked on Dr. Caruso who stood facing the surveillance mirrow. He turned, looked into her tired face; she nervously swallowed, and fell into his arms.

"Thanks . . . really thank you. What would we have done without you, Dr. Caruso?"

Henson only nodded; the man was a miracle worker, he thought. Jenny Lynn smiled, squeezing Caruso's arm.

An officer stepped into the entry of the hallway, nodded at Henson, and the detective touched Prissy's shoulder.

"Your pilot is ready, young lady."

Prissy weakly smiled, her stomach nauseous, as she released her arms from around Caruso's neck, and took the hand of the officer.

Henson twisted his torso and faced Jenny Lynn. "I understand you know Brian." See nodded. "Perhaps you can help us—come with me, please."

Henson opened the door to the interview room, and ushered Jenny Lynn ahead of his entry trailing behind, and pushed the door shut.

He politely pulled back a chair and Jenny Lynn sat down, her blue eyes gazing at the distant stare of Brian's deep, dark eyes. His face looked so pale, his eyes glazed over, as if he was in another world. Perhaps he was, she reasoned.

Jenny Lynn quietly spoke as she touched his hand. "Brian it's me, your old teacher."

Deep within the receptors of Brian's brain, a synapse signaled a memory, a pleasant experience that he'd stored there one lonely and destitute night, as he sat alone in his dormitory room. It was the night his uncle had informed him by phone that his funds were cutoff, the night he tendered his resignation for medical school, the same night he had decided to enter the Army with thoughts of picking up his medical studies while serving his country. Funny, he visualized, how his failing mind even recalled that night; it was his third year of medical school, and somehow he clearly remembered details. But that voice, that wonderful soft voice always soothed his psychotic emotions. His dry lips wiggled and cracked.

"Jenny Lynn, is it really you?" he whispered.

"I'm here, Brian."

He spoke with a quiver in his voice. "Tell me what to do. I'm not sure what I have done."

Tears rolled down Jenny Lynn's cheeks as she looked up at Henson, then at Robby, pain written on her pristine face.

Caruso placed his supportive hand on her shoulder. Henson swallowed; he had just witnessed an event that psychologists only dreamed.

"Ask him to sign the confession," Henson whispered, leaning over, while glancing up at his lawyer.

The lawyer nodded affirmatively. It was his best defense, perhaps his only defense.

Again Jenny Lynn touched Brian's hand. "Brian, the best defense you have is to voluntarily sign this confession," she advised, as Henson slowly slid the form under his eyes.

Brian gazed into the face of the only person he had ever trusted, clarity of mind that appeared only in her presence. He slowly, but deliberately, picked up the pen, and signed the form. He looked, for the last time, into the face of the woman he deeply admired and respected, painfully his cracked lips opened.

"Jenny Lynn," his voice faltered. "Dr. Kosaku, my uncle, paid me to steal Dr. John Cook's research, but I failed."

He faltered again, deeply sighing from a quivering chest.

"I swear to you that I did not kill your husband," he gasped.

Jenny Lynn's eyes closed. Tears flooded her face. She took his hand and held it against her cheek.

"I believe you, Brian."

Henson took Jenny Lynn by the arm as she stood, and led her out of the room, admiration registered on his face. Caruso proceeded behind them through the door into the hallway, and placed his arm around Jenny Lynn, winking at Henson.

Henson smiled uncommonly. "Dinner's on me wherever you say."

"I'll hold you to that promise," Caruso nodded.

As the couple walked down the hallway toward the front office, Henson consulted his mind. Could he believe that Latham did not kill John Cook? It was another bucket of worms thrust into this case. The thought reminded him of the fingerprints from Cook's safe; that was the missing piece of the puzzle. He made a mental note to check the forensics lab.

Chapter 37

A NURSE IN the cancer wing of Cook Medical Center reached over and gripped the receiver of the ringing desk phone, and then pressed it to her ear. The room was lined with hospital beds and several nurses scurrying around.

"Cancer Wing, how may I direct your call?"

"Judy, this is Jenny Lynn. Is Dr. Cook available?"

"One moment please," she replied and punched the intercom button.

As Jenny Lynn waited, Caruso reached over the desk and touched her hand. "It will be just find," he consoled.

Her face spread into a gracious smile, and she cupped her hand over the phone. "He's being paged."

Before he could answer, the phone clicked.

"Hi Mother."

"Hello, Chip. Wonder if you could have lunch with me and Dr. Caruso in the cafeteria."

"Say about 1:00," he suggested.

"That's great, see you then, and thank you, son."

Caruso stood and walked around the desk, and pulled Jenny Lynn into his arms. He planted a kiss on her lips, rubbed his broad hand up her spine.

She melted in his memorable arms.

***

The streetlights were lit in Boca Raton, Florida; a misty halo surrounded the glowing lights. Royal palm trees with long dropping fawns accented the corner of an expanding pharmaceutical company. The owner sat at his desk conceiving a plan of desperation, as he raised the phone receiver on his desk to a hairy ear, and punched a button on a handheld remote that shutoff a wall-mounted television. He had just watched a police alert on the capture of a serial killer identified as Brian Latham.

"Hello, I would like a reservation on the next flight departing for Japan," said his gravel voice, rasping like a file on steel.

He slid open the middle desk drawer and took out a Beretta handgun. "Yes, I'm holding . . . Tomorrow at 5 pm—yes, thank you. I'll take it . . . the name is Dr. Kosaku."

Kosaku closed the phone, and holstered the handgun strapped over his shoulder inside his coat. And then he stood and walked over to a safe hidden behind a van Gogh painting. He mused at the priceless painting for a brief moment. This famous artist had spent his last years in an asylum. His nephew was similarly crazy; however, it was doubtful that Brain would be sent to a mental home of the criminally insane; he'd killed too many people to be granted such mercy. He grunted and swung the hinged painting aside. His hairy fingers gripped the knurled knob of a wall safe: two reverse turns and right to a memorized number, back to zero. The steel safe door swung open; he shoved his fat hand inside, and gripped a metal box. He closed the safe door, and took the box to his desk as he thought for a brief moment. It was the logical decision, he would disappear. Then he strolled to a wine cabinet against the wall and poured a glass of vintage wine, and then slumped in his leather seat, his robust body squeezing between the chair arms.

After he swilled down a second glass of wine, neatly manicured fingers gripped the empty glass, and crashed it against the office door. Ragged slivers of broken glass squashed on the floor in random patterns. Trembling hands opened the metal box. He took out an envelope that contained fifty thousand shares of his company stock, including thirty thousand dollars in cash, and slid the envelope into his right inside pocket. He had previously prepared for this moment; a moment he considered would result from the nebulous mind of his neurotic nephew. He had already deposited two-million dollars in a Switzerland bank account, paid by Cook Medical Corporation through his resourceful partner for the purchase of his pharmaceutical company. A dismal thought coursed his mind: could he trust this partner?

He retrieved another glass, and emptied the wine bottle. He cherished this moment of silence.

***

Two mature women sat in a familiar café somewhere in the Bronx of New York. Jenny Lynn needed the advice of her old friend before she had lunch with Robby and Chip, although she really had no assurance that she'd take it. Still she trusted Dorothy Millhouse implicitly. They had been friends since she first arrived in New York, more years than she cared to remember. They had met at this very café on several occasions throughout the years. Dorothy by now had married her dance director, and he'd produced several Broadway shows with the prospect of a movie deal. Jenny Lynn reached across the tiny round marble table, and touched Dorothy's hand.

"Well, Dorothy, my dear, dear friend, the time has come. My old boyfriend is here, not only here, but he still loves me!" she confessed.

Dorothy sighed, tossed the end of her mink stole over her round shoulders. "Well it's quite clear to me what you should do, dahlin'."

Jenny Lynn slowly shook her head side-to-side. "Is it really that simple? I mean look at it logically: A man I loved twenty years ago comes back into my life. I have born his son without his knowledge and have married another man. Tell me how that is so simple," she exclaimed giddily!

Dorothy patted the hand of a woman she understood better than she herself, a woman who had suffered the pangs of womanhood, lost the one she truly loved, and married another. Why: Because the child demanded it.

"Everything is simple when you set your mind to do it, dahlin'. That's your problem, honey. You can't make a decision that stares you in the face—you're blinded by your own grief, blindness that affects not only you, but the happiness of several other people who love you."

Tiny tears nestled in the corners of Jenny Lynn's blue eyes. "Oh Dorothy, darling, how I wish it was all that clear to me." Her head bowed shamefully. "What must I do?" she exclaimed, pounding her fist on the table!

Dorothy stood, swung the mink stole over her shoulder. "Do, my sweet? I'll tell you what you must do," she barked, and bowed into her face. "You tell that dahlin' man he has a son," she said in her Mississippi drawl.

Jenny Lynn shook her head. She knew that Dorothy was sincere, had been ever since the day they became friends here in this very café. Until she met Dorothy, she had been alone for so many years that she was forced to make her own decisions. Reality had hardened her heart, and now that her first love had found her, she was desperate to keep him. Dorothy was right. She knew it in her scarred heart, but could she accept it?

Chapter 38

A THREESOME SAT in the Medical Center cafeteria, seated around a table, each holding a steaming cup of coffee. Caruso stared across the table into the hazel eyes of Dr. Chip Cook, a handsome young man that he had wanted to converse with for several reasons; a meeting that Jenny Lynn for some unknown motive had curiously evaded. Somehow this young physician reminded Caruso of his younger days in university. He'd opted for a doctorate in organic chemistry; because his military service had convinced him he had not the bedside manner of a medical doctor. But this young man had all the accolades. And somehow he felt immense pride for his splendid accomplishments in the field of cancer research, though he admitted much of his reasoning resulted from Chip being Jenny Lynn's son.

"Well, young man. I've heard so much about you, and I've looked forward to this meeting with some envy, I must admit," he blushed.

Chip seemed embarrassed, but sensed that in the presence of Caruso he somehow felt more at ease than he'd felt for a long time—perhaps fifteen years ago as a naïve middle-school student, when his mother dropped him off before going to the hospital.

"Likewise, sir, I think I heard your name more than once from my mother in my youth. Although I can't place the exact time, it was around the week that my father was killed."

Jenny Lynn touched her son's hand. "You jest, my boy. I don't remember any such thing," she smiled.

Chip only grinned, because he always yielded when his mother disavowed something he knew had a hint of truth. She was, and still is, the best thing in his life. He loved her dearly and only wanted the best for her. She had lived alone for so many years. And somehow he couldn't disagree with her fondness of this Dr. Caruso; he seemed a fine man, a man he strangely accepted as a friend.

Jenny Lynn changed the subject. "Chip, do you remember Brian Latham?"

He taxed his memory and finally found an image in the matrix of his mind. "Yes, I remember that he dropped out of medical school."

She nodded and took a sip of coffee, paused a moment, as she looked into his hazel eyes, the same color of Robby's eyes. "Brian is the man who killed Beverly Barrow."

The young medical physician sat stunned, his mind stirred by his discussions with Detective Henson. Yet he still felt guilty for how he had treated Beverly. He couldn't intelligently reply.

Caruso leaned forward. "Chip, we have Brian in custody. He is quite mad. He has confessed that he killed his only sister."

Chip sat speechless. Two killings, all related to his research; if only he might consult with his father, he thought. Sympathy welled in his throat, mingled with harbored guilt, and surfaced as rage, an emotion he had learned to control in the years of his medical schooling. The many times he had crashed in Beverly's apartment. And never once had he given her any thanks. Chip had no inkling of Beverly's infatuation with him, nor had he given any indication of romantic feelings for her; he never thought of her in that way.

Then it hit him!

He'd used her for his own purpose! Chip felt a sudden dirtiness, unfitness, distrust, a cad with no compassion. But alas, Beverly actually loved Brian, he supposed, and yet Brian had no viable love, either; he was driven by madness that stemmed from psychotic sources, a madness that he nursed with pills.

Jenny Lynn felt her motherly attachment to Chip's grief; should she add more, and then she remembered Dorothy's words. She told herself Chip must get it all behind him—all of it, but she was not ready to tell him the truth about his real father, despite Dorothy's advice! Was she as mad as Brain? Dorothy might agree; she knew her every emotion.

Caruso had waited for Chip's absorption of the bad news. He understood how he felt. When Jenny Lynn left University hospital in Jacksonville, Florida, he had returned to the hospital in search of her, to apologize, to do anything but lose her. She left with no forwarding address, no word of why she had gone away, or what he might have done—it had to be something he did, but what? Yes, he had absorbed similar bad news, and he was ready to end his search for happiness. If only Jenny Lynn would marry him, then he would retire. There, he had admitted it, if only in the secrecy of his heart.

"Chip, why don't you come with me and we'll visit Brad. I think your presence as a physician would be good medicine. Brad is conscious now," Caruso suggested.

Chip looked at his mother and saw agreement in her face, then he rethought his schedule; Debbie was not due for another treatment until tomorrow afternoon.

"Why yes, Dr. Caruso, that's very thoughtful— would you excuse us mother?"

176

Jenny Lynn replied with a broad smile. She had been remarkably released from the drudgery of revealing her secret. Rationalization was her nemesis, Dorothy her crutch.

***

Two men took the elevator and got off on the fifth floor. Chip and Caruso walked down the hall to Room 212. As Chip pushed open the door, he saw Henson resting in a chair. Brad was watching the football game on the overhead television. Chip immediately checked the patient's pulse, as Brad twisted his head toward Caruso. "I'm told I owe my life to your swift emergency treatment, Dr. Caruso."

Caruso grinned. "Prissy did the needle work for the direct transfusion. It was her blood that saved you, Brad."

"Yes, I owe you both my deep gratitude," he responded weakly.

Finally between brief chit-chats, Chip took the chart hung on the end of the bed.

Caruso quickly changed subjects with a question. "How is he, Dr. Cook?"

"He's alive and kicking, the best of news for any physician. Dr. Gordon Cash is taking good care of the Captain."

Henson's cellphone suddenly buzzed. He hurriedly fumbled in his pocket and released the electronic monster.

"Yes . . . when . . . the time . . . meet you there."

He closed his phone in deep thought. Could he pull his men off a high-profile case?

Brad rustled in his bed. "What is it?"

His gaze faltered. "Dr. Kosaku has flown the cope. He's purchased an airline ticket to Japan, departing in three hours."

Chip's head rotated and faced Henson. "That's the fish we were looking for, detective—the hook is set."

"Right you are, Chip. It's time to pull in the fish."

Brad rose on an elbow, dismissing the pain in his side, encouraged by the case. "Well what are you waiting for, gentlemen. You'll have the wrath of Prissy Shelly on your heads—get going."

Henson nodded. "Dr. Caruso, I could use a little help; seems I'm short of men."

"Why certainly."

Chip grabbed Henson's arm. "I bet I'm a better fisherman than you."

Henson smiled. "Come on, both of you, we'll take the police helicopter," he motioned with tobacco stained fingers, intrigued by his own uncommon action.

Chapter 39

DEEP DOWN IN a West Virginia abandoned coal mine, a column of geothermal steam pushed through the shale crevices, and released its mounting pressure into the bowels of the two-mile deep coal shaft. The tectonic plates had shifted some distant miles toward the west, and the magma layer was ominously oozing through the cracks.

In the early morning hours, as dew settled on the lawns, the earth's surface suddenly trembled. Animals and birds departed from the mountainous area toward the forests in the valley of upper New York. The stillness in the NYPD downtown jail suddenly vanished with a noisy shift in the concrete floor. The steel doors lifted from its hinges, the weight twisted the lock, and the released doors creakily fell open. The night guard in the front room lay on the floor unconscious; the result of a ceiling beam that fell, creased his head, and crashed through his desk.

Several prisoners seemed mystified, until one prisoner walked out of the cell and the others followed, except one inmate. This disturbed man sat on his bunk with his hands pressed against his head, as tormenting pain ran down his spine and exploded at the coccyx. He franticly stood, wildly screaming, as he stumbled out of the jail, piercing pain thrusting through the sciatic nervous system, and torturously lodging in lobes of his brain.

The streetlights had lost power, and Brian Latham sat dazed on a bench in the darkness outside the jail. His mind suddenly cleared in a moment of unusual clarity, and he focused on his uncle's pharmacy in Boca Rotan, Florida.

He curiously walked back into the jail with a plan haphazardly planted in his psychotic mind. He removed the outer clothes from the unconscious guard, and redressed him in his orange jumpsuit. Hurriedly he donned the trousers and shirt, although the shoes where a size too large, still he laced them. He took the Smith and Wesson handgun from the guard, stuffed it in his belt, and gripped a black leather jacket from a wall-mounted rack. He left the jail, as the dim light of the early morning lit the horizon in a haze.

He walked along the broken sidewalk, stealthily staying close to the building. The sparse crowds of nightlight were busily recovering from the tremors, and had not noticed his escape. Suddenly it dawned on his cluttered mind that he needed cash if he were to travel to Florida. He chose a storefront, and broke the glass in a rear window. Using his Army experience with mechanical devices, he jimmied the cash register drawer with a screwdriver he took from a janitor's closet. He found only fifty dollars, but reasoned it wasn't enough, and went into the back office. He pried open the lock on a file cabinet with the screwdriver, and found a metal box. It took only a second to break the tiny lock on the cash box. To his great surprise, he found a thousand dollars in twenty-dollar notes.

***

As Brian left through the same entrance that he'd entered, he heard the blare of distant police sirens crying in the misty morning. He raced into the rear area of the building, and disappeared in a dry ditch. The deep trench ran about five hundred yards and ended at a culvert. He saw a car lot beyond the culvert that bordered on a four-lane road, now empty and dark with loss of streetlights and cracked asphalt roads from the quake tremor.

Brian crouched between the cars and trucks until he surprisingly found a sedan that was not crushed by falling rocks, though it fortunately had an opened rear door. He gripped his throbbing head as he thought that some excited customer had left it open. Yet it was his way of escape, no more hiding in darkness, no more running from authorities, at last he would be free. The uncommon emotion was exhilarating. He reached over the front seat, unlocked the front door, and finally settled behind the steering wheel. It took only a few minutes until he removed a few wires beneath the dash and rerouted the battery power around the ignition to the starter motor.

The engine started with a spark from the twisted ignition wires.

Chapter 40

A LONELY HELICOPTER flew out to meet the late afternoon sun, while down below the busy life of New York City began anew, as construction crews labored furiously restoring the damaged streets and buildings. The governor had called out the National Guard, and the Army sent in its best engineers. The President called the governor and offered reduced interest loans.

Finally the epicenter tremors that emanated from the Virginias had ceased. Normal sounds once again echoed up and down the streets. Blaring horns, rearing mounted police horses, taxis waiting at the curbs, flags waving in the stirring breeze, pigeons were pecking the gravel, and traffic roaring down the freeways. The hustle and bustle of daily life was again on parade with a sea of humanity flowing from the communities to bask in the sun once more. They marched down the repaired sidewalks and into the parks, and patches of green crisscrossed by newly poured concrete. And life went about its merry way just as nothing had happened.

***

A robust man stood in line at the Orlando airport and waited to claim a reservation ticket to Japan. He carried a satchel in his broad left hand, a large Styrofoam cup of coffee in his right.

Hidden in the shadows between several tall marble columns, a disturbed man watched the ticket line. As Dr. Kosaku left the line with his boarding pass in his hand, he found a seat and sat reading an unfolded newspaper until time to board the flight. Brian Latham left his hiding place, and weaved through the columns until he was thirty feet from where his Uncle sat. Finally he stood near his uncle's seat. Suddenly his mind snapped, pain coursed down his spine, and forced him to his knees.

He screamed!

Kosaku stood startled, and faced the screaming man as his eyes popped wide open.

"I thought you were in jail!"

Brian placed his hand on the back of the seat and pulled himself to his feet, his eyes streaky red.

"You . . . you my dear uncle who reared me after my father and mother died in a rental house fire that you owned and refused to repair a broken gas line—you have caused me great pain, and I have given undeserved pain to a lot of innocent people."

Kosaku felt danger. "Now wait, Brian. You can't blame all this on me."

His glazed eyes blinked wildly. "I don't blame you—I challenge your morality, your humanness. You stole another man's research. You used me for your own purposes—you deserve to die," he screeched.

Brian's brain suddenly exploded with pulsating pain.

Kosaku seized the opportunity. He pulled a Berretta handgun from his coat holster. As he aimed the muzzle, a devious smile crossed his face.

A shot echoed up-and-down the marble halls.

Kosaku grabbed his chest and slumped to the floor in a pool of blood.

Henson rushed into airport lobby followed by Caruso and Cook, as crowds screamed, and stooped, some hid behind their luggage, others ran behind the columns. Henson kneeled by the dead body, and picked up the handgun with a pen slid through its trigger guard. He saw the startled crowds, stood, and waved his badge.

"Police, the excitement is over! Return as you were."

Several airport security guards rushed in and calmed the crowds. Henson instructed them to place the body in a body bag, and explained that he would take the prisoner. They agreed to transport the body his helicopter. Henson dropped the handgun into a plastic bag.

***

Out on the tarmac three men and a handcuffed prisoner walked toward a helicopter parked near the hanger. Chip closed his cellphone after he had made a call to his mother at Cook Medical Center. She was quite startled at Kosaku's attempt to shoot Brian. Henson informed his office on the helicopter radio and instructed his sergeant to see that the jail repair was completed before they arrived, and to alert the Coroner. Caruso and Chip sat in the back seat, Caruso's logical mind was busily considering if the case was closed or expanding. There were at least two more people involved somehow, Rufus Billings and this Mike Quincy. This case was still unsolved, too many suspects.

The flight back was quiet except for the sound of the helicopter rotors churning in the night air. Many questions stirred within the minds of the three men, but Brian sat in a daze, and Dr. Kosaku lay in a body bag behind the rear seat. Finally, Chip Cook voiced a message.

"Mother reports the earthquake has shut down the Roosevelt Bridge."

"Yeah?" quizzed Caruso. Yet the message seemed strange—an earthquake in New York? He thought perhaps it had something to do with the movement of tectonic plates, but he had no evidence, and it wasn't his expertise, just trivia that occupied his busy mind.

Henson considered Caruso's short response. "Seems a little out of character in these parts to me, too, but we also had some damage to the jail," Henson said, but his mind focused on his prisoner, and all the details he had to record in his report. More than that, he was concerned if this case was really solved. Other cases were piling up on his desk, but none as high-profile as this one. And there was his Captain, his need of evidence.

Chip's mind was on all the people who had died for the research on the cancer project. The deaths had mounted since his father began the study. Was it worth it? Then the image of little Debbie appeared in his mind.

"Yes, it is worth it—for Debbie if for no other reason," he suddenly blurted out aloud.

Astonished faces stared at each other. "Who is Debbie," they asked in chorus.

Chip smiled. "A little girl on the Cancer Wing back at Cook Medical."

***

As the helicopter approached New York, the increasing air traffic over the city told the group that the city was slowly recovering from the earthquake. Henson answered his suddenly buzzing cellphone. Finally he flipped closed the phone, and advised the pilot to reroute to another precinct because the jail repairs were more extensive than suspected.

The helicopter finally landed on the roof of the selected precinct at about eleven o'clock in the late morning. The lights in the building were still out, and emergency battery power dimly lit the hallway. As they walked down the steps to the elevators, they quickly realized the stairs were the only way to the police jail on the ground floor; elevators were inoperative.

The jail hallway, too, was lit by battery power, and a Sergeant met them at the door. He took charge of the prisoner, and placed him in a repaired cell, still other cells were not serviceable; the damage more extensive. Henson moved toward the office as he surveyed the damage, wagging his head side-to-side, not realizing the damage was so massive in this part of the city.

"Didn't expect this damage here?"

"It beats me, too, sir. All I know is the microwave towers are still okay, and cellphones are working."

"Well, is there any coffee around here?"

"Yes sir, Sergeant Mahoney made a pot on his mobile barbecue grill."

"Lead us to it."

He took the three men down the hall to a waiting area. Several cops were standing around the grill, not for food—they'd already fed themselves—the only wanted to get warm before going back out on the beat.

"Here you are, sir."

"Thanks. There is a body bag on helicopter would you see that the Coroner gets the body? Is there any word on the damage to Cook Medical Center?"

"Can do, sir. The latest word is Cook Medical is on emergency power like us," he said waving to a duty sergeant.

He whispered in the ear of the sergeant, and he left for the roof to pick up the body. One of the other cops spoke up.

"Sir, we just got a report about an hour ago that Cook Medical is operational, now."

Henson nodded. "Is the coroner available?"

He dropped his gaze at the detective, blinking his tired eyes; too many hours on the bench, he thought. "I called his office, but apparently the lines are still down."

Dr. Cook poured a cup of coffee realizing these men were victims of strenuous conditions, turning his head toward Dr. Caruso.

"Mother didn't mention any problems at the center. We have an emergency gasoline turbine generator that feeds the hospital with three phases of power, and we even sell the excess power to the city utilities."

Caruso poured a cup of black coffee. "That's good planning, Chip. My testing laboratory in Jacksonville, Florida, is a little less sophisticated. We have two gasoline generators that operate the equipment, and one on the lights."

Henson had different thoughts; among the more critical, were the test results on the fingerprint from Chip's lab ready. If not, he would send them to the FBI lab in Washington—he had to have answers?

Chapter 41

THE WEEKEND LOCATION that Henson had chosen for an outing with Dr. Caruso was on the Chesapeake Bay, quite a distance from New York City proper, but his mother owned a house on the beach. He'd invited Dr. Caruso to spend a weekend of peace and quiet; besides, the man deserved a vacation, if only for a weekend, the time needed for the fingerprints analyses. Dr. Caruso's invitational reply was surprising, not that he disagreed, but only if Jenny Lynn came, too. After hearing the story of their relationship twenty years ago and their recent reunion, Henson thought a weekend with his mother was a future wedding gift, although he hadn't mentioned this idea with either of his guests.

That evening, after a meal cooked by Mrs. Henson proved to be as smooth as the glassy water on the bay; exhilarating, a peaceful mix of bliss and solitude. Dr. Caruso sat with Henson on a long pier, each with a beer nestled in a hand. A pelican sat on a post overlooking the water and searched for its dinner. A flock of ducks silhouetted in a familiar "V" shape on the smooth water projected by the golden light of the setting sun, silently coursing overhead, interrupted by a distant honk or two.

Henson rustled in his chair and pointed at the pelican. "That old bird has been sitting on that post since I was a boy."

Caruso sipped his beer. "You must have had a good childhood."

"I was born in Rochester, my father ran a lobster boat from here; his boat is in that shed by the water over there," he said, gesturing toward the recently painted structure.

Caruso's old Navy memories surfaced. "Yeah, does the engine still function?"

"Purrs like a kitten, dad had it rebuilt before he died last year."

Caruso swiped the back of a hand across his mouth and crushed the empty beer can in his strong hands. "I'm sorry to hear about your father, but your mother surely can cook."

He released his lips from the beer can. "I want to thank you for bringing Jenny Lynn along, she's such a comfort to mother."

"I'm glad, too," he responded, and pointed the crushed beer can at the boathouse. "Say, wonder if we might take out the boat before it gets too late."

Henson's face beamed. "Capital idea. I'll just run to the house and advise mother, and pick up a couple of jackets. It gets chilly out on the water."

Caruso nodded. It wasn't a pleasure cruise that prompted his suggestion. Peter had called before he and Jenny Lynn had left for the drive to the Henson's house. Peter had discussed an investigation germane to Henson's case; apparently the case was not solved as everyone had hoped.

**

The boat was held out of the water by a gantry of chains powered by a three-horse electric motor; the barnacles had been removed in the recent summer. Henson punched the power button attached to a post, and a pulley system slowly lowered the boat into the water. Caruso and Henson stepped into the boat. Henson moved to the wheelhouse and cranked the diesel engine. Smoke belched from the dual exhaust, and the engine roared to life. He pushed a code on a remote and the boathouse doors swung open.

Henson turned to Caruso who leaned against the sidewalls. "Anyplace in particular you want to go?"

Caruso pointed. "What about that boatel café across the bay?"

He bobbed his head. "They serve a good steak, but it's a little gloomy inside."

"It's your boat."

"What the heck, I think I promised you a steak," he replied, and pushed the gas lever forward.

A breeze blew in from the Atlantic and both men raised the collars of their jackets. The air was brisk and breathable; cleared the sinuses, and filled the lungs with clean salty air. The approaching nightfall cloaked the boathouse in a blanket of mist, as colored lights lit the walkway on the boat docks ahead.

Henson reduced the speed and rotated the steering wheel in his alignment to a slip at the boat dock. He cut the engine as Caruso stood out on rear deck, and touched the dock with a boathook pole. Caruso jumped off and landed flatfooted on the dock with the bowline in his hand. He wrestled the line to a tie-off post.

Henson joined him on the boardwalk, and led him into the parking lot where the café front doors were located. A tall, slender man approached as they neared the door. Caruso turned to Henson.

"I'm afraid I've used you for a little covert action, Henson."

Henson smiled. "Why should I expect anything else from a man close to the President?"

"That's very generous, Henson. Like you to meet an old friend, Peter Meirs."

"He's that bean stalk walking toward us?"

190

"That's him, the only man who drinks more coffee than me."

Peter extended his hand and Caruso grasped his slender fingers. "Is this the detective you mentioned," Peter asked?

"Yes. Henson this is Peter Meirs, former assistant to the director of the CIA."

Henson gulped.

The man stood head and shoulders above Henson, had an unsmiling face that looked as if his muscles were glued. He pulled a cigarette from a sliver case and flashed a Zippo lighter with the Marine insignia on one side, the opposite side a relief of *semper fi,* and rolled a tiny wheel with his thumb. A flame the size of a Christmas candle lit his cigarette. He puffed several smoke rings that quickly evaporated in the breeze. Meirs extended a hand, clammy as a lobster; his face had the impression of a zombie.

"Henson," he nodded.

"Pleased to meet you Mr. Meirs," Henson replied as he took his hand.

Meirs removed his cigarette. "Peter, we go by first names."

Henson pointed to the entrance. "Well Peter, shall we go in, I need a drink."

***

Three men entered the marina café, and Henson led them to a familiar table that faced the bay. Henson sat facing the two government men and suddenly wondered what this meeting was about. It wasn't often that he had entertained two men that answered to the President of the United States.

A female waitress approached the table with pad in hand and a pen stuck in her hair. Her long hair was secured with a snood folded against the nape. She wore an apron tied around a wasp-thin waist and chewed on gum.

191

"What'll it be, Detective Henson?" she smacked.

Henson spoke up with his index finger waving. "I'll take a beer for starters, Margie–what would you guys want?"

Caruso ordered coffee, and Peter asked her to bring a pot.

She scribbled a few lines on a wrinkled pad, stuck the pencil behind her ear. "Be right back, gentlemen."

Henson sighed, decided he'd wait for the reinforcement of the beer before he asked a few obvious questions. But Caruso took advantage of the quietness.

"Peter, I think it advisable if you would enlighten our friend."

The retired government man nodded, and placed his lengthy forearms on the table. "Henson, the President often uses Caruso and me to trace cold cases. Caruso asked me to look into your case, and I have a photo that might interest you," he announced, as he pulled a file from his briefcase. Henson had not completely recovered from the shock, but his eyes focused on the photo, and his astonishment increased to the explosive level.

"How did you get this photo?" he barked, the only response his mind allowed under the circumstances.

"Taken two days ago with a camera operated by one of my former men," Peter casually replied.

Caruso took the photo from Henson's shaky hand. There were two men in the photo and Henson identified one as Mike Quincy, the other identity unknown.

"Well, Randolph," Caruso injected, "perhaps your case continues."

Henson swallowed. Randolph, he thought, no one had ever called him by that name except the

Captain. His mother had been talking about his childhood again, he surmised, as the coffee and beer arrived.

Peter sipped on his coffee, and slightly gestured his head toward the bar. "Don't look now, but the two characters in the photo are sitting over there—it seems they come in here quite often," he announced, having gathered this information from his informant.

Caruso smiled and stared across the way squinting at the unknown face, Quincy he had seen before. Henson sat aghast as he frantically peered at the two men. Peter puffed on his second cigarette; a cloud of smoke hovered over the table. A similar cloud swirled in Henson's mind, a startling confirmation that his case was far from being solved.

The waitress cancelled their thoughts as she approached the table. "Okay, guys if you want food just holler," she said between smacks of chewing gum.

Henson grabbed his beer and swilled a large gulp, as Peter poured two cups of steaming coffee.

Caruso touched Henson's hand. "What about one of those steaks you mentioned."

He smiled. "Great idea," he replied as he swallowed the third gulp of beer. "Mr. Meirs, how about you?"

"Rare steak, but no blood, please."

Henson sipped his beer, as Caruso opened his cellphone and punched a logo of Jenny Lynn's cellphone number.

Jenny Lynn sat with Henson's mother in the front parlor, and answered her cellphone. When she learned that they were having steaks, she knew everything was okay. It was always about Robby's stomach. She closed the cellphone and took a cup of tea handed by Henson's mother.

The silence around the table in the boathouse was suddenly pierced by the loud buzz of Peter's cellphone, as they sat waiting for the steaks. All eyes were upon the tall wiry man, who grunted twice, and closed his cellphone.

"It seems that the second man seated beside Mike Quincy is a cousin on his wife's side—does that mean anything?" Peter asked.

Henson nodded his head. "Yes, Alice Quincy had several cousins according to our investigation," he replied. Were all the members of this family neurotic, he thought. He swilled down the last of the beer contemplating the next surprise.

Chapter 42

CARUSO SAT AT a desk in Jenny Lynn's office while she was on duty at the Cook Medical Center. He considered how Chip would react if he were told what he'd discovered, neither was he sure what it meant. He decided it best not to broach this incomplete information tonight, because Chip was busy in the clinical trials. Brian Latham was in jail but a case would not be filed with the attorney general until after the fingerprints tests were available. In view of the latest information, perhaps that prospect was improving since Henson had called the Attorney General in Washington, against his Captain's advice; perhaps it would get him demoted, but he wanted the case closed, and Caruso and Meirs were his best bet. Henson had no political ambitions as did his Captain.

Caruso engaged the hospital's Wi-Fi connection with his cellphone, and searched the Billings family tree. After all that Henson had mentioned about Alice Quincy's family, it was his opportunity to validate his skepticism. To his surprise, the tiny screen suddenly flashed with a list of Billings' relatives, ranging from petty theft to robbery, yet firearms were neither involved nor current addresses supplied. He made a note for Henson to follow up the list. His cellphone suddenly rang and his thoughts vanished.

"Hello . . . Yes, Peter . . . yeah, thought we'd meet for coffee somewhere—oh, I see. Well thanks old buddy for the stakeout. We'll be in touch."

***

The hospital cafeteria was sparsely seated, and Caruso took a seat by the sunny window. Before he opened his briefcase, Jenny Lynn walked up and touched his shoulder.

His head turned. "Hello darling, won't you sit," he suggested.

As she sat, she took his hand. "Did you get much sleep after the long drive home last night?"

"I'm more of a night person these days."

"Yeah, know what you mean," she admitted. Somehow she found it easier to admit things to Robby than to Dorothy.

Caruso looked into her blue eyes and kissed her hand. "Henson has a photograph of two men seated together at that café where we ate steaks last night. There is a question about one of the men, more questions about them both, and their frequent meetings at the café, according to Peter's stakeout."

"Why don't you get us coffee and we'll discuss it," she wisely suggested, giving her time to visit her thoughts.

As Caruso arrived at Jenny Lynn's table with the coffee, he saw Henson's wave. He walked into the cafeteria and went through the line, poured a cup of coffee from an urn on the counter, and sat down beside Caruso, facing Jenny Lynn.

"Well, you two. Thought I'd join you for a moment, I'm on my way back to Midtown." Henson said.

Jenny Lynn flashed her blue eyes at Caruso. "We were just discussing a certain photograph," she replied giddily.

Henson casually remarked about their jovial behavior. "Have you guys been on a picnic, or perhaps had too much to drink on that trip to the Chesapeake Bay yesterday?"

Caruso's smile relaxed and he sipped a swallow of coffee. "Perhaps you brought that photograph."

Jenny Lynn drank her coffee quietly listening to the men discuss a subject more or less boring. She was more concerned about Brian, how his life was in shambles, probably he'd be executed. She asked herself many questions about this man; how he'd been so dedicated to the study of medicine, how circumstances beyond his control had stabbed him in the back. The sound of Henson's briefcase slamming on the table smashed her thoughts.

Detective Henson laid a photo on the table.

Jenny Lynn hovered over the photograph. Almost immediately her eyes spotted a fuzzy image.

"Got something?" Caruso blurted, peering at her wrinkled forehead.

She nodded. "There's a man in the shadows at the end of the counter—can't make it out," she stuttered as if she disbelieved her eyes.

Caruso stared at the fuzzy shadow in the background, Jenny Lynn's eagle eyes sending signals to her mind. She slumped into her seat, a million faces coursing through her mind.

"If you can enlarge that photo, maybe we can identify the man," she replied.

"Can you venture a guess?" Henson squirmed.

She evaded the question. She wanted a change of clothes before dinner tonight with Robby and Chip; yet deep in her mind she thought it time that she took Dorothy Millhouse's advice. Perhaps dinner tonight, she mused.

Henson opened his cellphone and emailed the photo to his office with a request to enlarge the fuzzy image.

Caruso piped up. "Why don't we meet you later Henson, it'll give you time to receive the enlarged photo."

He nodded affirmatively.

Chapter 43

JENNY LYNN ARRIVED at her home in the afternoon, the sun hung low on the distant horizon, long dark shadows silhouetted on the lawn. She assumed that dew had formed on the grass when she saw it glisten in the streetlights, then she remembered the lawn sprinklers had cutoff in the early morning. She chided herself for wasting time, the files at home could be very important, and she looked forward to dinner with Chip and Robby, yet she still had doubts about revealing Chip's biological father—Dorothy would be furious at her habit of procrastination. She parked in the driveway.

Jenny Lynn reached the concrete porch, navigated the three steps, and then fumbled in her purse for a ring of keys. Finally she inserted the door key in the massive lock. As she rotated the key and swung the door open, a man suddenly grabbed her arm from behind, twisted it behind her back, and unceremoniously shoved her through the doorway. He closed the 42-inch mahogany door, and released her arm with a warning.

"No screaming or you're dead," a raspy voice replied with a gesture of the muzzle of a firearm.

The intruder was of average height, wore a black knitted hood, and had an unusual smell she didn't recognize. Jenny Lynn's nose wrinkled at the musty odor, certainly not collogue, she thought. His shoes were canvas with herringbone soles, which she assumed by the wet footprints on the hardwood floors.

The man ripped a cord from a rewired 17th century lamp with Tiffany Art Nouveau glass and iridescent shade. Jenny Lynn cringed as the lamp rocked back-and-forth but the priceless antique finally settled on the glass-top Boulle end table. He grabbed her arms and pulled them behind her back, wrapped the cord around her hands in a figure-eight knot. Then he rushed directly to the office as if he'd been in the mansion before.

Jenny Lynn watched his shadow disappear around the corner, and twisted her hands in a seesaw motion in a desperate attempt to escape. Sweat lubricated the cord turns, one turn suddenly slipped over a palm, then two, and then she was free. She grabbed her purse from the floor and quickly looked toward the office, the coast was clear, and she tiptoed to the door. Gently, she turned the knob, ran down the steps, and jumped into her car.

She released the brake and placed the shift in neutral, and the vehicle noiselessly rolled down the driveway and into the street. She nervously inserted the key in the ignition and cranked the engine, shifted into drive, and pushed the accelerator to the floorboard. She roared off, her tires screeching, which left black tire prints on the pavement. She glazed through the rearview mirror and saw the man on the front porch with the knitted hood in his hand. She committed his face to memory.

Fortunately she'd left her cellphone on the seat, and quickly punched a preset code for Robby's cellphone as she raced through a red light. She heard the buzz of the cellphone, but it wasn't the phone a police car was behind her with his lights flashing.

"Darn," she whispered aloud as she pulled the car off the road and stopped.

Her fingers tapped nervously on the steering wheel as she peered through the rearview mirror. A policeman pranced to the side of the car, as she punched a button and the window lowered.

He bent down his tall frame. "Could I see your driving license please?"

She already had the plastic card available and handed it out the window. As the policeman took the card he heard a metallic sounding voice on the seat.

He removed his spectacles. "It's illegal to text while driving, you know—you ran through a red light, ma'am."

Jenny Lynn could no longer hold her rage. "Officer, I have just escaped from an intruder in my house," she barked as she gripped the cellphone and placed it against her ear.

The officer had stopped many women drivers before, but this excuse topped the pile.

"Robby, its Jenny Lynn, someone is in my house, I mean he bound my hands but I escaped."

"*Good grief. Henson is here now, we'll rush right over. You go to the Medical Center, I'll meet you there,*" a voice screamed in the tiny speaker, a sound the officer heard.

"I'm afraid I've been stopped by a policeman. Would you speak to him, please?" She handed the plastic nuisance to the officer.

Guardedly, but curious, he took the cellphone, surprised by her actions, his mind searching for a precedent. "This is Officer Johnston. Your lady just ran a red light . . . what! . . . who? . . . The President! . . . Yes sir!" He handed over the cellphone, "I'll escort you ma'am—where to?"

"Cook Medical Center, officer—quickly."

Chapter 44

THE MOON HUNG high and cast a blanket of eerie shadows creeping over the wide landscape sprinkled with two-million dollar stylish houses. The stars were obscured by the glare of streetlights that spread a misty veil over the subdivision. Out near Long Island Sound the hue of the horizon was a deep indigo as two police cars and a van pulled to the curb of a mansion. Three civilians, one a woman stepped from the first vehicle and rushed to the front door stomping the freshly fallen snow from their feet. Two policemen and two forensic technicians followed the trio in brisk step.

Jenny Lynn inserted her key but the door was unlocked. Dr. Caruso cautiously pushed opened the massive door, and Detective Henson followed the couple into the foyer, hand on his pistol. One policeman stood guard outside, and the other surveyed the premises. The two forensic guys navigated the steps, and walked through the door, each carried a black bag, and one had a Nikon camera with the strap slung over his neck.

The cameraman placed identification markers by two evidence sites and flashed a sequence of pictures, one of the twisted lamp cord another of the Tiffany lamp; he sealed the twisted cord in a sample bag as he surveyed the room. He moved to the doorway and placed a ruler beside the footprint on the floor, and then flashed several shots. The other technician dusted several smudges and placed a special plastic tape on a file drawer, then pealed it off, placed it into an evidence bag. It was a thumb print, the only tangible identifying evidence; the footprint was a more complex sample.

Outside a policeman had found a scarred windowsill where an instrument, probably a screwdriver, had attempted entry. A forensic camera man flashed several pictures.

When the technicians were finished, Henson said something to one man, and the forensic group promptly left.

Caruso stood beside Jenny Lynn, his arm around her waist. "What about John Cook's files, did the intruder take them?"

Jenny Lynn took a deep breath. "Don't think so, they are stored in the basement—that file cabinet stored here contains home files, tax records, and such."

Henson walked alongside a policeman toward the basement door, which the lady of the house had identified. The group descended the steps one-by-one, a distance of twenty-five steps downward. When they reached the basement floor and assembled around a stack of boxes, Henson opened his cellphone. He walked over to Jenny Lynn and showed her the enlarged picture of the fuzzy image of a second man in the picture that Peter Meirs had provided.

She studied the image but for a second. "I thought that was him, but I had to be sure."

Henson's face frowned. "Well who is he?"

She deeply sighed as she glanced at Robby. "It's Morris McKinley, a member of the Board of Directors."

Henson's eyes ballooned. "That's two suspicious members on the hospital board—who is running that hospital, a group freaks?" he barked annoyingly.

Jenny Lynn sat down on a stack of boxes. "I'm not quite sure; John never mixed business with his home life. But a wife can tell when something bother's her husband. I only know he and McKinley were once friends."

Yet she had not told the entire truth. McKinley had dated her on two occasions before John had proposed. He was a nice guy and had had dinner with her and John on several occasions.

The air was a bit stuffy, but Jenny Lynn's statement of Morris McKinley had opened new evidence for investigation, and Henson was anxious to get at it, to put it mildly. He decided the evidence demanded his exit, but Caruso advised his presence while they opened the boxes. He reluctantly agreed, and as they sifted through the boxes, Jenny went upstairs to change clothes. Obviously the dinner date was off and she took the opportunity to call Chip.

Dr. Caruso sliced the tape on one box with a penknife, and sniffed the familiar odor of sulfurized paper. Henson pulled a file from the box and laid it open on another closed box. He read a few pages and felt there was more information than Jenny Lynn could reveal. And so they waited for her return.

Finally Jenny Lynn returned with a pot of coffee and three cups. She found Robby with his back turned opening a box, with Henson on his tiptoes peering over his shoulder. Henson took one file and Caruso grabbed another, both determined to discover a lead that uncovered a motive.

Jenny Lynn interrupted, but the smell of coffee required no explanation, and they all sat on a box for a rest period. As they drank coffee, Jenny Lynn stared at a familiar box, and walked over to it. Both men sat their cups on a shelf and followed her. Robby opened the box with his penknife. Three pair of eyes read file-after-file. Finally Jenny Lynn raised her head with distant memories flashing before her eyes.

"Guys, I have something here."

Caruso dropped his file on a box and stood behind Jenny Lynn. Henson leaned against the stack of boxes with his legs crossed, his arms folded over his chest, and waited for a clue worthy of investigation.

"So, what have you found?" Caruso asked as he wiped the dust from his deep blue trousers.

She sighed deeply. "It seems that McKinley had a copy of the map, John records here. Apparently he had reservations about bringing him in as a partner."

Caruso's left eyebrow arched. "Partner in what?"

Jenny Lynn nodded. "John developed a small clinic into a major hospital. McKinley's bank financed the construction."

Henson digested the information. "There could be a motive here."

Caruso nodded his head, an index finger pressed to a cleft in his chin. "Then we will just have to find the proof—if it exists."

Henson thought how glib the remark, but he reserved his conclusion because he had seen this man accomplish miraculous things; it was simply good police work to follow every lead. Now he was even more anxious to get to his office.

Jenny Lynn closed the file and stood, gingerly stretched her back, then faced Caruso.

"I'm a little hungry, like to change into some suitable clothes, then you can take me out, just the two of us—I've already called Chip, the dinner is off."

Caruso nodded with a pleasant smile. He had just remembered his own hunger, but for him a hot cup of coffee alone with Jenny Lynn was exceedingly sufficient.

Henson took the file and walked to the stairway, turned with his hand on the base post, one foot on the first step. "I'll just keep this file; you two have fun—you deserve some time off."

As the couple followed Henson out the basement door into the hallway, Jenny Lynn whispered in Caruso's ear. "Dr. Rosenberg may have some answers."

He tightened his grip around her waist. "Who's he?"

She smiled. "It's a she. Dr. Rosie Rosenberg, chairperson of the Board of Directors."

Chapter 45

THE EVENING DRIFTED into a romantic hour as two former lovers sat at a corner table in a secluded restaurant, a fisherman's hangout near the water on Long Island Sound. The candlelight on the table suddenly flicked from a slight breeze as the front door opened, and a strange man entered and sat down at the bar. He ordered a beer, rubbed his hands together, and gazed around the room, curiosity on his unshaven face. He spotted the corner table and quickly turned his back, then stood and moved to a table in the shadows of an opposite corner.

Caruso placed his hand on Jenny Lynn's arm, a frown on his wrinkled face. "That man looks familiar somehow."

She nodded. "I think you're right, but who?"

Caruso wagged his head as fuzzy faces flashed across the retina like a kaleidoscope: negative search. Then he gazed at Jenny Lynn, a question stirring from the depths of his intelligent mind.

"Let's not waste this precious time; that face with come to me."

A waitress placed two plates on the table with steaming flounder, and a bottle of vintage wine. Caruso inhaled the smell recalling his days in Jacksonville, Florida, days when he had taken Jenny Lynn to the Turtle Restaurant on the Atlantic Coast, and how much fun they had enjoyed. His face glowed with the memory of long ago days.

Jenny Lynn spread a napkin in her lap and reached for a fork. "What are you thinking; I've seen that expression before?"

Robby poured two glasses of wine and sipped a swallow. "You remember the Turtle Restaurant?"

She chuckled. "The night you fed the seagulls, or the night you forgot to pick me up?"

A sheepish grin flushed his face. "You're never going to let me forget it, are you?"

She placed her hand on his deeply tanned hand, the same hand that had caressed her so many years ago.

"What's to forgive, I experienced it, and I want to experience it again," she said shyly.

Caruso's heart leaped into his throat. A question stored there among the scars, among the lonely nights that now pushed to the tip of his tongue. "Jenny Lynn," he stammered nervously, "Jenny Lynn, he repeated with a dry swallowed . . . would you marry me?" There, he had finally said it, and the tension finally released.

She sat astounded, a flood of tears cascading down her cheeks, his words reverberating in her heart, sweeping across the scars that had punished her during countless sleepless nights. Yet her joy had no words, only the desire to hold him always. Deep in the recesses of secrecy, she had closed the subject that she most feared—discovery of the son that was his, but she loved this man and would not let him go.

She stretched out her hand and lay it aside his tanned cheek. "Yes, Robby, yes, I will marry you—I will," she sobbed relentlessly.

***

The strange man seated at the bar cloaked by a raised collar stood, pulled his cap over his eyes, and strolled toward the door. His approaching shadow swept away

Caruso's attention from the pinnacle of ecstasy, and he suddenly stood.

"Wait up there, fella—I want to talk to you."

The man bolted out the door.

Caruso turned to his fiancée. "That man has been following us."

"Great Scott, be careful, Robby!" Her heart smiled; John was always running off to the hospital, and Robby, he was still a kid at heart.

"Right!"

He ran through the door and searched the dark parking lot, his eyes those of an eagle looking for prey. A moving shadow caught his eyes, and Caruso rushed toward a spot near the water. He looked both ways, and suddenly the muzzle of a pistol grazed his head. Caruso abruptly turned, grabbed the man's wrist; the pistol fired, a bullet thudded into the man's hip, and he crashed to the sand, rolled to a kneeling position as Caruso heard the hammer of the pistol cock. He surged into the assailant's stomach with his head and forced him to the coastline. They wrestled in the waves crashing on the beach. Quickly, Robby socked his chin with an uppercut, and then straddled his prostrate body, as he grabbed the pistol that had fallen on the sand.

Jenny Lynn raced out to the beach yelling into her cellphone as the owner of the restaurant followed her, a towel wrapped around his robust waist flapping in the breeze.

The owner assisted Caruso to his feet, and brushed the sand from his wet clothes. Jenny Lynn fell into his arms. "Are you hurt, darling?"

Caruso filled his lungs with fresh salty air. "Afraid that guy is."

"Oh darling, I'm so glad you aren't hurt—I called 911, an ambulance is on the way."

The owner seemed remorseful. "I'm deeply sorry, sir."

Caruso placed his hand on the owner's shoulder. "Quite all right, wasn't your fault. Do you know this man, I mean have seen him here before?"

He gazed into the face of the unconscious man for a moment, his hand on his chin, and then a bushy eyebrow arched above a deeply green eye. "I think I have seen him here before, but with two other gentlemen," he replied with Irish eyes smiling.

Caruso nodded, his mind racing with questions. "Listen, it's important that you recognize those faces again?"

The shrill of a police car shattered the conversation, it's red and blue lights rotating like a strobe light. Henson bounded to the shoreline, followed by two paramedics with a stretcher.

"Well, Caruso what do we have here–thought you two went to dinner, is this your desert?" he chuckled.

Caruso's face spread into a toothy smile. "This man followed us into the restaurant," he pointed. "Meet the owner, mister . . ."

"O'Malley, Roger O'Malley," he smiled. He was in his late sixties, round and full-bodied with unkempt silver hair and thick untangled beard. He would have easily passed as Santa Claus at a Christmas party.

Caruso shook his hand, "Pleased to meet you, Roger. This is Detective Henson."

Henson tipped his forehead with an index finger. "Mr. O'Malley. Wonder if you might come down to the precinct, like to show you a few pictures."

He smiled as the breeze rustled his reddish hair. "Glad, too–how about tomorrow around noon."

Henson nodded, "See you then, thanks for your participation," he said, handing him a business card.

His rosy face lit up like a jack-o-lantern. "Listen, gentlemen. I love this country, unless we all chip in, crime and corruption with takeover."

Caruso took his hand in a handshake. "When you come tomorrow I'd like to talk with you, O'Malley."

Jenny Lynn pressed her nose on Caruso's cheek. "You may be busy honey, you promised me breakfast."

It was her reminder that he'd promised to marry her. But would he trust her after he learned that he had a son, she thought. She had chided herself over and over again when Robby ran out to follow this man. What if he was killed, she couldn't bear the thought. She must and would tell him the secret. If she didn't, someone else would, it suddenly seized her mind.

O'Malley smiled. "I'll find you, sir—you and the lady have a good breakfast."

Again, Caruso smiled. "Bring you wife along, too, O'Malley."

O'Malley nodded. It was good that he had met people like these, he thought. He'd spent countless hours among fisherman, derelicts, and single women looking for a handout. Yes, he would go to the precinct tomorrow, and his wife, too. He wanted to know more about this man and his fiancée.

Henson gazed down and realized the man was conscious, and stooped. "An ambulance is coming, bud—hold on; you have some explaining to do."

Only one thought impressed Henson's mind. If Caruso stayed another week he'd solve the entire case, and he hoped he would stay—he liked the guy.

Suddenly the sounds of an ambulance crossing into the parking lot stole their attention. Two paramedics carefully placed the wounded man on a stretcher and hooked up a glucose drip. They hooked an oxygen mask over the patient's mouth, and lifted the stretcher as the wheel frames locked. They rolled the unit to the rear of the ambulance, collapsed the wheels, and shoved it forward until it clicked securely in its locking bracket.

Chapter 46

A WOUNDED MAN lay in a hospital bed on the second floor with an IV stuck in a vein above his wrist, glucose feeding into his bloodstream. Dr. Michael Brown had removed a thirty-two caliber slung from his hip, and his condition was good except for loss of blood, which a blood transfusion had solved. But in the man's mind he realized there was a worse problem, because he had overheard a nurse say that Detective Henson was due any moment. He decided he'd hold Henson to his promise for his testimony. The door to his room opened and his thoughts vanished, but not his concern. Whatever happened, he was ready to tell everything he knew, but he felt no elation about what he must do.

Henson entered the room, and the nurse left the bedside; the detective's eyes followed the door as it closed. Henson turned and focused his squinted eyes on the man in the bed. He tossed his hat on the chest-of-draws, just missing a box of vinyl gloves, which reminded him that he hadn't received that report from Prissy's fingerprint analysis. He finally sat down in a chair beside a nervous patient.

"You realize that you are about to take the rap for several deaths," he snarled.

The patient managed a weak smile while his mind digested the gravity of the statement. Finally he released his nervousness with a deep sigh.

"We are a close-knit family."

Henson's left eyebrow arched. "Well that remains to be seen. I want to know exactly what you know–who are you, is a good start."

His head dropped, and stared directly at his tanned hands contrasted on the white sheets, thoughts rushing thorough his mine.

"My name is Rufus Billings, Brian Latham is my cousin."

"This Dr. Kosaku is Latham's uncle?"

"Yeah, on his mother's side."

"So, tell me what you know."

"Brian called me and we met, wanted me to open a safe."

"And what do you know about safes."

"I worked for Wells Fargo ten years ago and we delivered money kept in safes, and often placed money in client's safes."

"I see, so you can open just any safe?"

"Not the safe Brian wanted me to open. It was hardened steel of a special process with entrance only through a special card–of course it could be opened with explosives, but that wasn't reasonable.

Rufus rotated his head, an attempt to shield Henson's cutting eyes, the quilt had returned but remorse flooded his thoughts.

"I also placed the bomb in Dr. Cook's car," he confessed.

Henson's head twisted, facts rushing through the filters of his mind. Perhaps all the cousins wore goatskin gloves, he thought; wonder if Prissy had isolated the different breeds, was there any chance that technology might identify which cousin wore the gloves in question. Why did he ever think this case was solved?

"So you masterminded that bomb, too!" he suddenly exclaimed.

Rufus defended his confession. "Yes, but Brian supplied the C-4."

"Well," Henson snarled. "You have quite a resourceful family."

Rufus considered his crime, the detective's deniable attitude. "Listen, you promised me immunity if I cooperated."

Henson nodded. "That's true, and I'll recommend your testimony to the judge. Now, what can you tell me about Mike Quincy?"

He dropped his gaze; a pain in his wounded hip interrupted his thoughts as he studied for a moment.

"When Mike married Alice, the family approved because he was a good businessman, and he ran the goat farm profitably."

"Goats?"

"We had some cattle early on, but the prices drastically dropped. When Mike took over the business he introduced goat-skin products, especially cashmere gloves. He bought two pair of Kashmir goats and started a herd. Mike established contracts with two French firms, three in Britain, and four out west."

Henson crossed a leg as his cellphone buzzed, and he flipped it open. "Henson here," he replied as his eyeballs suddenly swelled when he heard the first few words.

For a chilling moment the room froze icy still and foreboding. Rufus lay traumatized as a nightmare of thoughts overloaded his frenzied mind. What was this call all about?

The message was from Prissy Shelly, and Henson finally closed the phone. The fingerprint was still unknown, no match in all the databanks of the police department, the CIA, or the FBI. And then he remembered the third man in the photo given by Peter Meirs that was in his briefcase. He closed the phone and looked directly into Rufus' eyes.

"You're not finished with Quincy, tell me more," Henson barked.

Rufus rubbed the ache in his leg, searching his mind. "All I can tell you is that Mike Quincy was a relative of a banker in Holland who came to this country through Ellis Island. He is probably the nicest guy of the family."

Henson's eyes reflected on his thoughts. Perhaps there was finally a connection; a honeycomb of cousins and relatives. Mike Quincy stood beside Rufus in that photograph that Peter Meirs gave him. There were many clues but none pointed to John Cook's killer, and Prissy had not identified the hair found on Cook's sleeve, or the fingerprints in the vault of Cook's lab.

Chapter 47

CHIP RETURNED TO his laboratory in the basement of the Cook Medical Center with the intention of working late. He had not heard from Dr. Caruso or detective Henson. He was too preoccupied with his many absences from the project, and financial ruin was staring in his face; the ten million was scarcely enough for this study. There was a feeling of helplessness, almost fear in his mind as the hour grew late. He worked late sifting through the financial deficits; apparently he had not watched the funds carefully enough. He remembered that Marcy had briefed him on the growing deficit. It seemed that he depended on Marcy more and more. Suddenly he shuddered that he might regard her like he had treated Beverly.

He sat back in his chair and rubbed his tired eyes. No longer were there facts, only misguided wishes. They were so many dying people with cancer in the study. The therapy results were extremely effective, yet he was running out of funds. Why must there be a price tag when cancer was a major killer? Yet he saw himself being swept on a tidal wave racing toward the rocks of financial ruin.

His depression was interrupted by an aide who entered his lab office and motioned toward the phone on his desk.

"You have a call, doctor, a Morris McKinley of City Bank."

He forced his composure into rigid calmness, and lifted the receiver. "Hello, Morris, I take it you don't have enough hours in the day either."

"Right you are," McKinley came back. "I thought you might like to know I have a lead on your expensive project."

"You'll have to explain."

"That pharmaceutical company the Board of Directors voted to purchase."

"Oh yes, the company Rosie mentioned at the Board Meeting."

"That's right. It's a company located in Boca Raton, Florida. My bank holds the mortgage."

Chip ran the city name through his mind, and thought that he remembered something about the place, but who had told him? Oh rats, he chided.

"So, have you got any idea when this facility will be available?"

"Probably late November," he replied. "I'm going by the place this weekend. There are a few items on the square footage and setbacks that the bank requires. How would you like to go with me?"

Chip consulted his mind on the weekend schedule and couldn't recall a conflict. Yet he belatedly remembered that Clyde Nevins, his old friend, may know something about the place. Perhaps he was the one who had told him about the pharmaceutical company, he thought, but couldn't remember. These late hours and all the financial details were taking its toll on his memory. And then he curiously relaxed as he thought of Marcy Curtis.

"Okay. Where shall we meet?" Chip asked.

"I could pick you up," he ventured.

His mind rebutted. "No, I might want to take someone with me," he replied.

218

"That's fine," McKinley hedged. "I should arrive by noon Saturday. The facility is located on the Main Street, you can't miss it."

"See you there, and thanks Morris."

As Chip hung the phone, he noticed a note on his computer screen that he had a recent e-mail. Curiously he clicked on his e-mail site, and discovered an e-mail had indeed arrived during the conversation with McKinley. He opened the message, quickly noticing an attachment. It was from Detective Henson with a subject title: *A photo taken at a boattel café across the Chesapeake Bay.* Now his curiosity almost choked in his throat as he opened the attachment.

His fatigued eyes squinted at the image of two men. An enlargement of a man in the shadows astounded the core of his being! His eyes enlarged, the mouth dropped, struck with amazement, shocked as if by a thousand volts.

"Morris McKinley" he exclaimed aloud.

Questions swamped his mind.

Thoughts erratically raced through the synapses of his mind, and he briefly wondered if Dr. Kosaku and McKinley had any relationship besides both being members of the Board of Directors. And suddenly an agonizing suspicion gripped his heart. What was the relationship of McKinley and his father? He needed facts; that much he had learned from Dr. Caruso, societal facts were no less important that medical facts.

Chapter 48

THE EARLY MORNING air reeked with the smell of the fall season, though it was only a bright sunny Friday morning, snow still covered the lawns in upper New York. The nostalgic smell reminded him of football games, tailgate parties, and long nights around a warm fireplace roasting chestnuts and marshmallows. It was about six o'clock, and Chip had already telephoned Clyde Nevins, and he was to meet him at the airport for a flight to Boca Rotan. He'd promised McKinley that he'd be at the pharmaceutical facility on Saturday noon, and if they got there a few hours beforehand, he and Clyde had a chance to investigate the premises. Marcy had agreed to watch the lab and follow the needs of the clinical trial, that is, if we had dinner on Sunday night.

Clyde finally arrived at the airport and approached Chip in the lobby cafeteria where he said he'd wait. As he walked up to the table, he dumped his thoughts. "We had better be prepared for most anything—I heard that Dr. Kosaku was killed in an attempt to escape to Japan."

Chip considered his comment only a moment. "Yes, I heard—remember I told you that people would kill for a cancer cure?"

Clyde squeezed the lobe of an ear. He'd do anything for Chip, not only was he his best friend, his formula had given great promise of life to his little daughter, Debbie. And somehow Chip seemed preoccupied as he watched him take his cellphone from his pocket.

Chip flipped open his cellphone and punched a preset key. The circuits buzzed; a call-waiting message. "Mother, this is Chip. Clyde Nevins and I are flying to Baton Raton at seven o'clock this morning for a meeting with Morris McKinley. We should be back early Sunday morning. I love you."

Clyde looked at the surprised expression on Chip's face as he closed the cellphone. A question formed in Clyde's mind.

"Well, what did she say?"

Chip dropped his gaze. "Left a message— thought it best to let someone know where we're going."

"Smart move."

He nodded. "Let's wait in the lobby for the flight."

***

The flight gates were overflowing with college kids going home for Thanksgiving. Fortunately Clyde had more experience traveling and he had made reservations for a direct flight to Miami. They would rent a car and drive to Boca Raton.

A light flashed on an overhead monitor, and their flight was on time. They hustled to the boarding gate and entered the lines at the ticket agent counter. Clyde gave their names and presented his boarding passes.

Finally they followed a line of boarders up the covered walk into the door of a 747 parked on the tarmac. The two men selected the seat numbers printed on the boarding pass; they found their seats located near the midsection. The flight attendant walked the aisles, and warned for all seats to be placed upright and seatbelts buckled.

When the Boeing marvel was airborne, Clyde ordered himself a beer and Chip a cola. This non-stop flight offered time for a chat. Margaret was greatly pleased at Debbie's progress, and he wanted Chip to know their gratitude. Yet, it seemed that Chip had other things on his mind, as he opened his briefcase and presented the photo he'd received in an e-mail from Henson.

Clyde swilled a swallow of beer. "This guy in the enlargement, who is he?"

Chip sat his cola in the can-hole on the arm of his seat and pointed to the photo. "That's Morris McKinley."

"McKinley, isn't he a board member?"

Chip sipped on the cola as he nodded affirmatively. "Yes, and that makes two suspicious guys on the Board—one is now dead."

Clyde's eyes squinted almost shut. He wanted to say something that gave Chip hope. He pulled his reading spectacles from his shirt pocket, and settled them firmly on the end of his nose. His eyes peered over the frames and he quickly shoved the spectacles to the bridge of his nose with an index finger. Suddenly his mind surged with a memory.

"Say, Morris is the guy I've seen at Dr. Kosaku's facility."

Chip munched on a complimentary chip and swallowed a gulp of coke.

"We had better be discreet when we meet McKinley; there is so much we don't know."

He gripped his beer can. "Yeah, ain't that the truth," he blurted out," unconcerned with his poor English.

Chip mulled over Clyde's response. Something was amiss, he could smell it. Perhaps the search of the Boca Raton facility would identify the missing puzzle pieces.

***

When they arrived at Boca Raton, Clyde went to rent a car, and Chip stopped by a food shop and purchased two coffees and a newspaper. While they waited for the car, Clyde made his excuse for a trip to the restroom. As he crossed the aisle to the stairway he noticed a man stepping out of a taxi through the outside glass door. Somehow he looked familiar but he couldn't focus on the distant face.

Clyde stood facing a mirror in the restroom washing his hands his mind buzzing: who was the in the taxi, he thought. It didn't occur to him that it might be McKinley. He dried his hands and tossed the paper towel in the trashcan, and burst out of the door. As he rounded the corner he suddenly saw the image seated at the adjacent boarding station. Clyde increased his steps, almost ran, and sat down beside Chip.

"See that man seated over there with a blue jacket?" he asked gasping for breath.

Chip closed his newspaper and squinted across the aisle. "That's Quincy."

"Yes, that's him. I wonder if he is waiting to see McKinley, too."

Clyde realized now how despondent Chip had become, the tone of his voice, and his lack of excitement. His mind quickly retraced the events of several meetings and focused on the day that he had first clarified Chip's confusion about the Boca Rotan chemical laboratory, how it was being converted into a pharmaceutical company. And when the Cook Hospital Board of Directors had announced its purchase of a laboratory, it became clear to him that the purchase was Kosaku's company. It was an enigma that he hoped might unravel tomorrow morning. Yes, he understood Chip's attitude, and now he was thankful that Chip had invited him on this trip. He vowed that

Chip would not get hurt–emotionally or physically, he was too valuable in research.

"Should we approach Quincy," Clyde wondered?

Chip thought but a moment. "No, let's wait and let the chips fall where they may."

Clyde thought it was a good decision.

Chapter 49

THE SUN SHONE brightly even at half past nine on Saturday morning, when Clyde and Chip arrived at the pharmaceutical company. Fortunately Clyde knew exactly where the facility was located, and they wasted no time finding the plant. Clyde drove a rental sedan up the driveway and parked in the back parking lot. Since the sale of the laboratory was pending, Dr. Kosaku had laid-off everyone but three department directors. He had prepaid them a sizeable bonus to remain and supervise the remodeling construction. Apparently the construction crew was not working today because the equipment was idle and only one car was in the parking lot.

The two men entered the back door but no one was in the building. Clyde curiously retraced his steps outside, and approached the single car in the parking lot. He touched the hood, reached under the frontend, and felt the radiator. The engine was stone cold. Apparently it belonged to one of the construction crew, or maybe a supervisor had left it overnight.

When Clyde returned, he found Chip in the front office sitting at a computer. He had already opened several files when Clyde peered over his shoulder.

"Found anything interesting?"

"Not yet."

Suddenly a file opened when he clicked on an unusual logo. Before they read the title of the document, a car stopped in the front parking lot. Chip hurriedly clicked out of the document and shutdown the computer, while Clyde peeped out the window.

"It's McKinley and Quincy," he whispered.

Chip stood and glided to the window. "They haven't seen our car, let's hide in the rear."

They entered the supply room located in a hall directly behind the secretarial office. The front door slammed shut, and the two men entered the front door noisily chatting as they walked past the supply room into the first office. Clyde suddenly left the supply room, and tip-toed to the open office door. He raised his digital cellphone and snapped a non-flash photo, and then quickly returned to the supply room.

Chip winked, and motioned his hand for Clyde to follow as they quietly exited through the side door into the back parking lot.

"Good work. That photo will give Henson photographic evidence. Why don't we keep our appointment, its eleven-thirty," he said, glancing at his wristwatch.

Clyde nodded, and they marched to their sedan. They slammed both doors for dramatic effect, and strode toward the front door. Chip stomped noisily into the room so that McKinley heard them. They walked into the front lobby and Chip called out: "McKinley!"

The two men in the office glanced at each other, and McKinley motioned Quincy to exit through the presidential door in the back wall of the office. Quincy had left the hospital in New York and had flown to Roca Raton; McKinley owed him money and he came to collect. His hip still bothered him, and he limped toward the door. Once Quincy was gone, McKinley answered the call. "Here, Chip. You're a little early."

McKinley entered the front lobby with his hand outstretched. He shook Chip's hand and faced Clyde, rolling his eyes from the tip of his head to his toes.

"Haven't I seen you around here before?"

Clyde bobbed his head. "Perhaps, I'm in pharmaceutics and I regularly call on this laboratory."

"I see," he replied and faced Chip. "Welcome to your new facility, Chip. There will be three processing lines when completed, a clean room, a large production lab with temperature control, plus offices and restrooms. The shipping/receiving building will complete the construction."

"Sounds great, any questions for me on the sale closing," Chip smugly replied.

"No. I think everything is in-order after I turn in a few measurements," he responded, a gleam in his eyes. "Say, have you two had lunch?"

Clyde's facial expression signaled that he was hungry but his motive was to secure more time for Chip to question this guy. "Could I pick up the tab?"

"No, it's on the bank. Would you follow me in your car?"

***

A sedan pulled into the driveway of a quaint little restaurant near the ocean and parked in the side parking lot. A warm breeze rustled the fronds of Royal palm trees that majestically lined the streets. Already the city was alive with tourists, construction workers, and concrete trucks whizzing past the little restaurant. As Chip and Clyde stepped from their car, McKinley opened the door of his Mercedes-Benz parked near the door.

They followed McKinley into the lowly lit café. A shapely dressed waitress ushered them to a table facing the ocean. As they waited for their orders, Chip watched the seagulls soaring overhead and a few pelicans perched on posts extended out of the water. A wooden pier projected about forty feet from a patio outside the window that faced the Gulf.

Chip released his gaze from the window. "What do you make of the death of Dr. Kosaku?"

McKinley raised his arms above a steaming platter, and spread a napkin in his lap.

"Quite extraordinary," he replied after a swallow of iced tea. "I heard the board was considering his removal."

"On what grounds," Chip asked, simply to keep the question open.

McKinley released his fork. "It seems that funds have been misused on the purchase agreement of your facility."

"What misused purchases?"

McKinley finished chewing and swallowed. "He was over-pricing purchases of supplies."

Clyde laid his fork on the edge of his platter, and wiped his mouth with a napkin. "Dr. Kosaku purchased several units of Inositol from my company."

McKinley's face frowned not from the sting of hot coffee but the probing reply. "Unusual wouldn't you say, Chip?"

"Detective Henson thinks so," Chip replied, and swallowed a bolus of steak and potatoes.

McKinley sensed that he had best end the line of discussion before he implicated himself. "How is the research going, Chip. Rosie seems quite enthused— just yesterday she asked the board to extend the ten-million dollar limit to fifteen million."

"Oh?"

"Yes. And the board approved the request."

Chip sat flabbergasted. "That's great—we could use the extra dollars."

Clyde couldn't resist mentioning his young daughter's progress in the clinical test. "I want to thank your bank for financing the program; the program has saved my daughter's life."

McKinley was struck by the remark. Somehow he wished things were different. John Cook would be proud of his son, and he himself certainly was not proud of what he'd done. Then his heart sank; he never had a son, never wanted to marry after Jenny Lynn had married John.

Chip responded as he wondered what was in McKinley's vague stare. "Debbie is responding admirably in the program. Yet I don't find any correlation in young patients versus older patients with the same cancer."

McKinley broke his gaze. "I'm afraid I'll have to end this enjoyable meeting, if I intend to get those measurements to the bank. You will excuse me," he said as he took the ticket, and strode to the register by the door.

As McKinley walked away, Chip sipped his coffee in deep thought.

Clyde laid his napkin on the platter. "He's worried about something, perhaps the Quincy connection."

Chip released his thoughts with a blank stare. "So am I, Clyde, so am I."

"I guess we had better get to the airport if you intend to get back for dinner with Marcy," Clyde advised.

Chapter 50

MARY SHAW WARMLY greeted Dr. Chip Cook, Dr. Robert Caruso, and detective Randolph Henson in the offices of the Board of Directors of Cook Medical Center, New York City. Chip had called Mary yesterday when he had arrived back in New York. He made the appointment, and then kept his date for dinner out with Marcy. Mary was always glad to see him and was excited that he had made the appointment, and so was Dr. Rosenberg. So many wonderful things had transpired in the board meetings, but Dr. Kosaku's death was not one of them.

Chip kissed Mary lightly on both her rosy cheeks as they hugged. He had known Mary when he was but a teenager and she had been extremely helpful in achieving the grant for the clinical study from the Board.

"How marvelous to see you again, Mary," said Chip, releasing her tiny arms.

Her wide smile warmed the entire room. "My little teenage feller, it's good to see how you've followed in your father's footsteps—how is the trial progressing?"

"Quite well, thanks to the Board for increasing the grant. How is Dr. Rosenberg?"

Mary dropped her smiled, consulted her calendar. "She asked about you only last week," she replied and paused. "You know, she wondered if you might call her, seems she has some personal information for you."

"Wonderful. Is she in her office today?"

Mary clicked the computer mouse. "About two o'clock–there's a Board Meeting at four o'clock."

Detective Henson rustled in his chair. "Do you have a picture of the board members that I could take with me?"

A frown creased Mary's forehead. "Yes, but could I ask why, or maybe I should ask what's this meeting about?"

Dr. Caruso rocked forward in his seat. "You no doubt know about Dr. Kosaku's death, I suspect."

Mary's left eye elevated, yet she had known Kosaku's involvement was suspected, and his death was all over the news. "And you have more incriminating information about the Board, I suspect?"

Chip touched her freckled hand. "Mary, we need background information on McKinley." As he looked into her brown eyes he saw a familiar caution. "Perhaps you had rather we speak with Dr. Rosenberg as the Chairman of the Board?"

Her head nodded. "Yes, Chip. This is out of my purview, but here's a brochure containing pictures of all the board members, their titles and corporate positions," she replied pulling a file from here desk drawer.

Henson stood. "Ms. Shaw. There are pending criminal charges but I'm willing to keep it quiet as possible."

She extended her aging hand. "Thank you detective Henson. I'm sure the Board will be grateful."

***

Four o'clock rolled around sooner than expected, as Dr. Caruso rustled in his seat after closing his cellphone. He and Henson had discussed the facts uncovered by each of their searches, both waiting for Chip's report on the meeting with Dr. Rosenberg. They sat in a coffee shop across the street from the hospital waiting for Chip. The two men had mutually agreed that Rosenberg was fond of the young physician, and despite Chip's persistence, they insisted that he keep the appointment alone; they were an albatross around his neck.

Mary had expected Chip's early arrival because he was always punctual, and had placed his appointment on the calendar after Rosenberg had called her office. Chip sat in Rosie's office and read a report on his clinical trial that he'd found on Mary's desk. His eyes fluttered and closed from lack of sleep, then opened when he heard the slamming door in the outer office. He placed the report back on the desk, and sat down with his arms crossed in his lap as he shook the cobwebs from his head.

He heard Dr. Rosenberg's voice conversing with Mary. Rosie opened her office door as Chip stood waiting.

"Why Chip, it's so wonderful to see you, again. I've been meaning to call you about the progress of your trial, the reports are a bit sketchy, and I notice we may have underestimated the price tag. Now what's own your mind, how can I help you?"

Chip nodded with his best smile. "I'd like to thank you personally for the wonderful increase in our grant for the clinical trials. We are all excited by the progress, and Marcy is doing a fabulous job in my absence," he replied and bowed his head slumping back into his seat. "Rosie, I'm afraid I'm flying under deceptive colors."

She sat. "Your deception is noted, let's get to the reason you are here."

"I must ask about Morris McKinley—I'm afraid he doesn't grasp the realities of this study." It was the best argument he dared to present.

She stared into his young face but a moment, her mind racing back twenty years.

"You know Chip, what I'm about to tell you is not for publication. I should have told you earlier but I wasn't sure you'd understand—I don't completely understand myself, even now."

As Chip sat with anticipation on his face, Dr. Rosenberg inserted a key into the lock of the mid-drawer of her desk, and retrieved a file, aged by time and frequent use. She slowly placed the file on the desktop as if it were the Holy Grail. She gently fanned several pages, and finally laid the file open at a specific page, and raised her head somberly.

"What I am about to disclose is based on my respect for your father, Chip. I have suspected for some time that McKinley was envious of John's research, and after John's death, I began to accumulate this file."

She spun the file around, and Chip leaned over and read several paragraphs as his mind accumulated several questions; the first concerned Dr. Kosaku, and another was the banker McKinley.

The fuzzy memory of an association of his father and McKinley had not seemed important to him as a teen. Rosenberg watched his expression as she poured a glass of water from the decanter, and offered Chip a glass, but he declined. She raised the glass to her lips, and then sat it on the desk.

"I suppose you have your reasons for investigating McKinley?" she ventured.

"Yes," he replied and looked into her unsuspecting face. "Detective Henson needs all the facts to complete his investigation on Dr. Kosaku's untimely death."

She nodded, accustom to gathering facts for her files. "I understand that this Brian Latham shot Dr. Kosaku."

"Yes, but the entire case is highly complex, involving several misfit cousins."

From her years of association with Chip she somehow wasn't convinced; a deeper question harbored in his young mind.

"Come now, Chip—what's the real reason?"

Chip rocked back in his seat for a pensive moment, captivated by Rosie's prying eyes. And then he leaned forward, and placed his elbows on the desk, both hands clasped together.

"Well, since you asked," he paused for a pregnant moment. "I think McKinley killed my father, not Kosaku!"

"What?" she screamed as she stormed to her feet! Yet Chip expressed no reaction. He'd only spoken the truth revealed by the facts.

She wagged her regal head as she opened a desk draw, and took out a bottle of bourbon. The sturdy chairwoman of the Board, a woman who had loved John Cook though she had respected his wishes, poured a glass and swilled it down, smacked her lips as she formed a rebuttal.

"I never thought McKinley capable of murder," she suddenly surmised. "Yet we suspected Kosaku was involved, he had issues with your father, something about disqualification of his nephew from med school."

Chip read between the words of her remarks. "Rosie, Brian Latham is Kosaku's nephew."

Her eyes ballooned. She slowly released her glass, facts and faces, coursing through her mind.

Chip sat stoic, logging every image in the memory cells of his cortex as he waited impatiently for Rosie's reply.

Rosie poured another glass of bourbon, swilled a deep gulp, and rocked back in her chair. "Well, that does explain a few discrepancies in Kosaku's actions."

She brought her hands together finger-to-finger beneath her powdered chin.

"I must admit that McKinley has been in my sights almost as long as Kosaku, but I couldn't fit the pieces together, that is, until last week when McKinley took a week off from the Board."

Chip sat antsy. It was nearly impossible to hold his tongue. Yet his mind was in overdrive. The Kosaku family tree was truly a twisted branch; but what about McKinley—who was the guy? Finally he looked into Rosie's eyes.

"I don't remember any details about McKinley's association with my father—it's all a blur."

Rosenberg eyes gazed into nothingness, her thoughts hidden in the deep fringes of her mind, sealed there as she grieved over John Cook's death. Her mind finally returned restlessly to reality, but the information had been lock there for several years. She swallowed; the pain in her throat was numbed by alcohol.

"You were only a toddler, Chip when your father and McKinley were involved in securing a map of a trail leading into the bush at the foot of Mount Kilimanjaro."

His eyes beamed. "I don't know where it originated, although my father did have a map." His facial expression disclosed the thought process calculating in his mind. "Yet, I do remember a conversation between my mother and father about a safari into East Africa."

She rocked forward and placed her elbows on the desk, her mind now more relaxed. "It's my understanding that your father and McKinley discovered a map during a trip into the Kenyan back country. An old explorer found the map and your father offered to buy it, but the request was refused; he didn't want to sell. After years of investigation, I received a report just last month that McKinley had stolen the map.

She faltered; the truth of McKinley's immoral behavior suddenly crushed her words. She poured another glass of bourbon, brought the glass to her lips, the sweet smell intoxicated her thoughts.

"After they returned to America, McKinley revealed the map to John, but didn't tell him how he'd acquired it. When John challenged him, McKinley gave him the map declaring a partnership. Your father confronted McKinley, who declined to reveal how he'd acquired the map. Both men had hair-trigger tempers, and they fought. McKinley received a black eye and a broken arm."

Chip sat aghast. "I never heard that story before."

Rosie poured a glass of water from a decanter on her desk, soberness increased her recollection.

"There's more. Soon after their return from Kilimanjaro with the map after a week of looking for a hidden hut without success, John disagreed with filing a partnership agreement, and McKinley threatened John with a demand of half of any proceeds from his research. John refuted his loyalty to the Hippocratic Oath, and accused McKinley of violating everything he stood for. McKinley scoffed and renewed his threat. They never spoke again."

Chip sighed, his index finger rubbing the deep cleft in his chin. He had not known about this relationship, nor had his mother as far as he knew. He was so overcome by the disclosure, so seized by disbelief, that he couldn't speak. It seemed clear that McKinley had a strong motive for murder, and it fit the assessment he had made without all the facts.

Rosie sat the glass of water on her desk, and stood. "There is one thing you can do for me, Chip. Mary tells me that detective Henson may make an arrest. I hope we can keep this under wraps, our stock would take a beating."

Chip raised his head. "I owe you more than that, Rosie. You can rest assured that detective Henson will be discrete," he promised as he stood.

"And your Dr. Caruso, how will her react?"

"Frankly, I don't know, but he's an honest man."

Her eyebrow arched. It was all she expected. She walked around the desk and placed her hands on Chip's shoulders.

"Chip, my darling, no matter what happens, you stay true to the clinical study. It was your father's wish that a reasonable cure could be found," she said, tears blurring her eyes; a deep seated memory straining at her heartstrings.

Before Chip replied, Rosie pulled him into her arms with a motherly hug. Now he understood that she visualized him as his father. How she must have loved him before his mother came on the scene, he thought. Life has no inhibitions—it is the people who make the choices—good or bad.

Henson stood when Chip walked in the coffee shop, and met him at the door.

"Well, what did she tell you?"

Chip grinned. "Could you wait just a moment, I could use something to drink."

Caruso understood Randolph's anxiousness, and went over to the urn and got three coffees. He placed the cups on the table and sat. Even before he reached for his cup, Randolph popped a question.

"Okay, what's on this broad's mind?"

Chip may have been offended, but he was beginning to understand Henson. His Captain must really be a strong boss. Before he answered the question he took a deep swallow of coffee as he smiled at Caruso.

"First of all, she gave me some very inside information that she said must be exchanged for keeping the Medical Center out of the conversation."

Caruso grinned when he saw Henson's face.

Chapter 51

JENNY LYNN QUIETLY considered Chip's earlier telecom; especially the information from Rosie. She was piqued by the fact that Rosie had dated John before she came on the scene. And somehow she was relieved, although she never had expected McKinley's involvement.   John had not mentioned McKinley's relationship, that is, not as Chip had revealed. It was so like John not to worry her needlessly, and she was confident that Henson would be discreet with the information.   After her shower, she phoned Robby and asked him to schedule a meeting with Henson.   Too many details were needlessly scrambled and needed clarification.   Chip promised to meet her in about an hour at the Medical Center. She planned a rendezvous in the cafeteria, that is, if her hair cooperated.

Finally she pulled on a pair of slacks and buttoned a matching blouse, when her cellphone buzzed. She stumbled on tip-toes to the bureau and sat, laid her hairbrush on the glass top, and flipped open the phone.

"Yes."

"Hi Mom, I'm in a taxi headed for the Medical Center. Wonder if we you could get Henson to meet with us, too."

"Yes son. I have already arranged it. How are you feeling today?"

"Tired, but I want to get this McKinley matter settled. It's eating at my guts."

She frowned, a mother's concern. "Don't worry, son. We've got the evidence, just be patient."

He deeply sighed. "I'm running low on patience, Mom. Action—I want action!"

She forced a smile. "I understand, son. Let's discuss it at the cafeteria."

Chip squirmed in his seat as horns blared past the taxi, everyone in a hurry, but his mother was always the calming force of his life.

"Thanks Mom. I have always depended on you. I love you."

Jenny Lynn's blue eyes flushed with moisture. "I love you, too, son."

***

The time for her guests' arrival had past; neither Henson nor Chip had arrived yet and Jenny Lynn sat famished; somehow, nervousness always kindled her appetite. She decided to get some food, stood, waltzed to the cafeteria line, and took a tray. The cafeteria served breakfast after the noon hour and through midnight, and she took two eggs over light, hash browns, and a cup of coffee. She paid with a credit card, and took a seat at a faraway table facing the entrance to the cafeteria. As she sat down, she suddenly remembered she had forgotten the cutlery and a napkin. When she returned with the items, she heard Chip talking in the hallway. As she sat down, both Chip and Henson entered the cafeteria.

"Hi Mom, sorry I'm a little late—New York nightlife slows down the traffic."

Henson noticed the steaming cup of coffee beside Jenny Lynn's plate, and detoured to the food line. He slid a piece of coconut pie on his tray and moved down to the coffee urn. Chip entered the line behind him and already had scrambled eggs, and sausage piled on his tray.

Jenny Lynn moved her tray, not wanting to sit between the two babbling men. Despite her efforts, Chip sat on one side of his mother, and Henson sat on the other. Each reached for their coffee cups as their arms crossed under Jenny Lynn's nose. The sat with their hands around the warm cups like two trained monkeys obeying their master.

Jenny Lynn sighed and opened the conversation. "Well Chip, what's this about your guts being eaten?"

A smile crossed the face of Henson. Chip sat his cup on the table. "I anticipate that we could bring McKinley in as a material witness," he replied, glancing at his mother's confident face, and then swapped his gaze at Henson. "How long can we hold him?"

Henson released his fork, and wiped his mouth with a napkin. "Depends on whether he demands a lawyer."

Jenny Lynn sipped her coffee quietly, the steam rising into her face. "McKinley is too smart to demand a lawyer. I think he will answer questions but not candidly."

Chip's anger stirred. "I want the truth, mother."

"Just what truth do you think McKinley is hiding?" Henson injected before she could reply, but Jenny Lynn's motherly instinct deferred.

Chip pushed back in his chair, the tension in his legs had reacted with his thoughts. "You know, Prissy just might prove that he killed my father," he charged flatly.

Henson frowned. "That's a long stretch, and he is not going to admit that he's a murderer."

Chip sipped a deep swallow of coffee, hoping the caffeine soothed his nerves. "What about baiting another hook?"

Jenny Lynn had listen to a great amount of trivia in her experience, but this conversation was going nowhere.

"Listen guys. We have evidence, circumstantial, but evidence with which you should confront McKinley— let him answer your questions, and stop your conjecturing. The map, his jealousy, Rosie's investigation, and what about this Mike Quincy—he should have information, too—that picture that Clyde took of them both at the laboratory opens up a flood of more questions. Bring in McKinsey, that's my answer."

The air was suddenly still, no rattling cups, no slurping coffee, only stares.

Henson finished the last bite of pie, and gripped his coffee cup. "That's good, she's right you know, Chip. We can interview Quincy and McKinley in separate rooms, and then replay the information against them."

Chip sighed. "Is that really legal?"

Henson swallowed a gulp of coffee. "Yes, we have sufficient information to hold him forty-eight hours. Often a sign of guilt arises when they demand a lawyer," he injected.

Chip bowed his head, digesting Henson's assessment. "You are the legal arm, Henson. Do what you have to do, but don't let McKinley get away," he cautioned with narrowed eyes.

Henson stared into the distance, recalling his past dealings with Chip. He was an intelligent young man, honest, and sincere, but he had no inkling of the criminal mind; how it worked, how it made decisions, the limits of sanity, victory, and defeat.

When the meeting was over, Henson issued a pickup authorization for both McKinley and Quincy.

Chapter 52

FRIDAY AFTERNOON ARRIVED and patrol officers of the NYPD police department had brought in Mike Quincy and Morris McKinley. Henson had successfully jumped through the hoops of the legal requirements laid down by the department lawyer, and it only took twenty four hours to arrange the interviews. The two men were taken to separate interrogation rooms. Jenny Lynn, Rosie Rothschild, Dr. Caruso, and Chip Cook all stood in the narrow hallway viewing the proceedings though the window/mirror.

Henson stepped into the room where Quincy sat, and tossed a file on the desk, and then pulled out a chair, straddled it like a saddle. He stared cunningly into a troubled face for what seemed an eternity. Finally he opened the file and laid out a photo.

"This is Rufus Billings—that's you Quincy behind the counter," he declared looking directly into his eyes. "What were you talking about?"

Quincy squirmed in his seat, an impression of shame and discomfort. Yet he sat mute.

Henson pulled out another photo of Dr. Kosaku's body. "Did you authorize this killing," he asked, his eyes glaring into in his face.

Mute: more squirming without comment.

Henson suddenly pounded his fist on the table shouting, "Somebody will pay for this murder—it could be you!" he exclaimed, pointing an index finger directly into his face like a dart approaching the bull's-eye.

Quincy's face suddenly erupted.

His emblazoned face shone livid. "Brian was a maniac–you know that!" he barked.

"So you *can* talk. What is your relationship with McKinley?"

Quincy's eyes enlarged as he wrung his hands, an expression of guilt crossed his face. "I'm just a go-between, they call me the dealer."

"Who calls you the dealer?"

"Brian and Rufus."

"Who were you dealing with?"

His face visibly tensed. "I can't answer that–it was only a voice on the phone."

A look of anger flushed Henson's face as he opened the file again, and pulled out a photo of the exhumed body of Dr. John Cook. "Do you know who this is?"

His eyes closed as he turned his head, visions of Alice's dead body seized his tongue.

Henson looked at him intensely. "This man was murdered, we suspect you did it," he charged recklessly.

His eyes enlarged beneath a wrinkled forehead, torment streaked across his face. His eyes nearly popped from their sockets. Rage released his tongue.

"No! No! I only received a plastic card from Brian and placed it in a locker at Grand Central as instructed." Quincy bowed his weighty head and began to sob profusely. He slowly removed his hand covering his mouth, rubbed his fist under a sniffing nose. "Really that's all I know," he sniveled. "When Brian killed my wife, I lost all hope, cared about nothing, suspended the tanning business, but I needed money to pay the bills. Rufus gave me a telephone number–the rest you know."

Henson closed the interrogation door behind his exit, and instructed two officers to take Quincy to a safe-house, the man had suffered enough, he thought. He strode to the window/mirror, facing the viewers behind the panel.

"Now let's see what McKinley will tell us."

Jenny Lynn formed a question as Henson left the interrogation room and strolled to the area behind the mirror.

"Wasn't goat hair found on John's sleeve?" she puzzled.

Caruso had remained silent until now. "I don't think Brian killed John."

The silence forced the air out of the room.

Chip's left eyebrow elevated. "Well, I think I would wait for the forensics analysis on the fingerprints."

Jenny Lynn didn't agree. "Explain, Robby."

"I believe Brian would have told you, Jenny Lynn, if he killed John, you saw his face when you persuaded him to sign that confession."

Henson gripped his index finger and thumb around his unshaven chin. "That's a little thin, Robby."

"Let's just assume that I'm right. If Quincy didn't kill John, then that leaves McKinley."

Henson had hoped Caruso might add something useful to the discussion, but this was child-like deduction. "How did you arrive at that far-out conclusion?"

Caruso smiled. "I just put myself in McKinley's place. He had ample opportunity to get a pair of those goatskin gloves, and he knew Brian wore them, then it follows that any evidence he left would incriminate Brian, not himself. He could be the go-between Rufus and McKinley alluded to."

Henson smiled. His faith in Caruso was unblemished. "Not bad Caruso—not bad at all. But can we prove it?"

Henson removed a file from his briefcase and went into the second interrogation room with two cups of coffee. As he opened the door, he saw McKinley clipping his nails nonchalantly. He dropped the file on the desk, sat the coffees on the table, pulled out a chair, straddled it, and sat down.

McKinley only smiled confidently as he dragged over a cup of the coffee. He wrapped his fingers around the warm cup; he knew every question Henson was about to ask him, and had an answer ready.

Henson returned his smug smile with a glaring stare.

The silence was deafening, the air electric.

The two men sat staring into each other's face across the table like two sparing boxers. Finally McKinley rustled and broke his stare.

"I came here under by my own free will, thought I could help the case, so go ahead, ask your questions," he barked, as he swilled down the entire cup of coffee.

Henson opened the file. He pulled out the picture of Cook's exhumed body. "Do you recognize this body?"

He sat the empty cup on the table. "Why should I," he snarled.

Henson's eyes stared like daggers. "It's John Cook! You killed him!"

McKinley bounced to his feet like a Mexican jumping bean. "I'll sue you for defamation of character."

The air suddenly stilled like a storm brewing.

"Sit down! Here's my cellphone, call your lawyer."

McKinley regained his composure with a sardonic smile, suddenly laughing, and a chilling sound that seemed to refrigerate the room.

He took the phone, wiped the smirk from his face, and punched in a code. Eyes exchange glances, as the number finally connected. McKinley turned his back with a raised eyebrow and whispered into the tiny speaker, his open palm wrapped around his mouth.

While his back was turned, Henson thrust a ballpoint through the handle of the empty coffee cup, and exited the room.

He walked the few steps along a narrow hallway to the mirror room, and extended the cup to Prissy, who stood with Caruso, Jenny Lynn, and Chip.

"Put on your plastic gloves, Prissy. This fingerprint may finally identify the killer of John Cook," he said, glancing at Jenny Lynn.

Prissy smiled as she turned to the gentle woman beside her. "Don't worry, Ms. Cook. We have some strong evidence here."

Henson's face spread into a rare smile. He felt this evidence would finally close this case. "Call me soon as you process this fingerprint—perhaps you can match those other prints, too."

"Sure thing, it shouldn't take more than a couple of hours."

As Prissy left the hallway, a lawyer strolled up to Henson. "Are you detective Henson?"

"That's me, and you must be McKinley's lawyer."

The Harvard graduate stood dressed in a black, pin-striped suit, his robust body anchored on highly polished alligator-leather shoes. He shifted his sizeable weight, switched his briefcase to the left hand, and extended the right hand. "Gilbert Southerland."

Henson released the clammy handshake and pointed to a door. "Your client is through that door, Mr. Southerland."

He nodded his cone-shaped head, a blanket of curly black hair, and walked through the door; an officer stood guard, as he passed. Henson issued a few instructions, and faced the gathering behind the mirror. "Would you people follow me—like to buy you a cup of coffee."

Caruso smiled. Chip mused. Jenny Lynn laced her slender arm around Robby's muscled arm and gripped his hand.

Chapter 53

A BRIGHT SUN flooded the office of detective Henson at the NYPD Police station, perhaps an omen signifying the report due from the forensics lab actually solved the murder case, which had taken too much of Henson's time, according to his Captain. Out in the lobby a female officer served hot coffee to Henson's guests, an event that had occurred rarely. But if the report was positive, he planned a dinner out for his guests.

Henson's cellphone buzzed. He flipped open the plastic nuisance and quickly pressed it against his ear. "Henson speaking."

The words were like the soft music from the strings of a golden harp. Henson slumped into his seat; a pressing weight had finally lifted off his shoulders. He mused the cherished moment. Yet still the past few weeks gnawed at his ego: No case in his files had taken so much manpower with so many clues without degrading to a cold file. He closed the cellphone.

***

Prissy ran up the stairs only about in the afternoon. In her briefcase she carried the report on the fingerprint analysis. She had left the Rochester forensics lab and hustled into Henson's office, huffing and puffing, and tossed the report on his desk.

"Frame it," she gasped.

Henson grabbed the report, patted Prissy on the shoulder, and proudly walked out into the lobby. All faces turned to the noise of his approaching footsteps and Prissy's high-pitched voice.

"McKinley killed your husband, Jenny Lynn. Dr. Caruso, you were right." Henson reported.

She sank into the chair; the news immediately relieved a mountain of depression from her lonely heart, but disbelief still accused her mind: McKinley seemed to be such a kind man.

Caruso sat down beside her, and placed an arm around her waist. "I'm deeply sorry, honey."

She laid her hand on his bronze-colored wrist. At least she had Robby, and now she felt confident to tell him that he had a son. The barriers were finally down, all inhibitions unchained.

Chip read every word in the report, traced every groove in the fingerprint photos. Not only had McKinley killed his father, he had also taken samples of inositol from his safe. The jumbled pieces fitted together like a jigsaw puzzle—a nightmare of disjointed clues had melted into to one picture: McKinley! Finally it was over, and now the clinical trial took precedence. "The clinical trial," his mind repeated. He had forgotten one thing, his dinner date with Marcy. Quickly, he checked his cellphone messages, and found an email from Marcy. She had rescheduled their appointment for the weekend; she had too many case files to peruse before the final phase of clinical tests. Her excuse touched his emotions. He owed her more than his guilt dared admit.

Henson left his conversation with his Captain who had cheerfully accepted his report. and authorized McKinley's arrest. He stood in a chair and shouted above the chatter: "Listen, you guys. Dinner is on me."

***

A four-door sedan pulled into the parking lot of Caruso's favorite Restaurant on Staten Island. It wasn't Henson's first choice but he owed Caruso a great deal, and Jenny Lynn, too–the gentle lady was responsible for his mother's recovery, and Chip; why this young man, was a miracle worker in his field of medicine. All was well with his world, and besides he was hungry.

When they walked into the lobby, O'Malley and his wife met them with infectious smiles. After they shook everyone's hand and offered several greetings, the robust owner escorted them to a large table in the backroom where executives met on weekends.

O'Malley served his best wine. He sat a plate of warm biscuits in the center of the table, and went to the kitchen until they all knew what to order. Henson poured the wine and Jenny Lynn passed the plate of biscuits as they each scanned the menu. Henson decided on a rare steak, Chip selected a New York steak, and Jenny Lynn liked shrimp and scallops. Caruso's mouth drooled for a good lobster dinner yet he somehow made no decision, still undecided. The waitress took their orders as they munched on biscuits and wine while she waited for Caruso's order. Jenny Lynn chided him for his indecision.

O'Malley suddenly pushed open the kitchen door with an extended elbow as he balanced a steaming plate of lobster, and placed it on the table before Caruso. Every eye focused on the huge Maine lobster, gently garnished with lemon and butter. Caruso looked into O'Malley's gleaming eyes.

O'Malley draped a towel over his arm and poured refills of wine, smiling into Caruso's hazel eyes.

Caruso smiled. "How did you know I wanted lobster?"

"I can spot a lobster lover from ten feet," he grinned.

"Well you surely read my mind," Caruso graciously replied.

O'Malley threw back his head, robustly laughing, his tummy shook like a bowl of jelly until his apron strap unloosed. As he tied his apron strings, the group all responded with clapping.

Two waiters came in with a large platter of cornbread and placed it in the center of the table, beside the plate of biscuits. They sat steaming plates before each smiling face. When they arrived at Caruso's seat, a waiter sat a plate of New York strip steak beside the plate of lobster.

"Jenny Lynn says you need this," O'Malley responded lyrically.

Caruso beamed as he cracked the shell of the seafood prize. Jenny Lynn smiled with pleasant memories; it had always been about Robby's stomach, and she remembered just how much she loved him. The quietness was interrupted by the sound of cutlery cutting into selected foods, and an occasional tinkle of a wine glass.

Henson finally wiped his mouth with a napkin and sat his wine glass on the table. "Chip, when will your product start production?"

Chip toyed with his fork, and slumped back in his seat. "Not until the last phase of the clinical trials are completed," he replied with misgiving thoughts of his frequent absence from the trials, his mind twisted around Marcy's dedication.

"When can we expect that?" Henson rebutted.

"Probably six to eight months," Chip responded from his memory of reading Marcy's files. Somehow he missed her smile, he suddenly reminisced.

Jenny Lynn bit into a buttered biscuit and placed it on her plate, gently wiped her mouth with a cloth napkin. "There will be a review board and then the

FDA will have to approve the product–takes a lot time to put a new drug on the market."

Henson pushed back his chair, his stomach already processing the nutrients. The steak was so tender it almost melted in his mouth. He'd finished dinner first, probably because he had not ordered a potato. As he laid the napkin on the table his cellphone buzzed.

"Henson," he answered with the phone pressed against an ear. As he listened his eyes squinted, a million questions swirled in his mind. "When," he asked abruptly? Again his eyes squinted beneath a wrinkled forehead, his head slowly wagging. "I'll be there in thirty minutes–better still, send the helicopter here . . . O'Malley's restaurant on Staten Island." He closed the phone, gazing into space.

The silence was eerie, the questions harbored seriously.

"Dr. Caruso would you mind coming with me. McKinley has escaped," Henson requested.

Eyes popped open, confusion sealed their speech.

Caruso realized it was not *how* he had escaped, but *where* he was headed.

Chapter 54

THE NIGHT WAS dark, the moon obscured by inclement weather. Yet it was just the cover that Quincy needed to sneak McKinley to his getaway vehicle parked in the alley. He had visited the police station soon after Henson had released him from the safe-house, ready to claim the reward promised long ago, when he decided to cooperate with McKinley. He had been arrested which set the reward into motion, and Quincy carefully made plans for the escape. Clandestine activity was not his thing, but the incentive of $50,000 was too inviting for refusal.

The last time McKinley and Dr. Kosaku had met, they devised a plan of escape if either of them required assistance. The plan included a flight to Switzerland. And since Quincy had been loyal as a go-between, they had agreed he was trustworthy. There was another $100,000 in a Swiss bank, and sizeable cash stored in the lab at Baca Raton. It was a last resort plan, but lucrative for Quincy, now that he was caught in the breach. It was blood money; a small payment for the death of his wife, Alice. The thought chastised the memory of her gentle love—oh how he missed her.

McKinley had not readily recognized Quincy when he came into the jail dressed in a police uniform. Quincy had watched the officers milling around in the area for the opportunity of a stolen cell key, and when an officer laid the ring of keys on the desk, his moment came. Fortunately the officer left the cellblock and joined a group of officers in the lobby. Quincy snatched the keys and dashed into the hallway, and quickly unlocked McKinley's cell. He rushed to his bunk and took him by the arm with a finger pressed over his lips. When McKinley realized the escape plan was underway he stood. Quincy crunched up his blanket in an elongated mass with a pillow crammed at the headboard hoping it provided the extended time needed for escape. He placed the ring of keys back where he took them.

They moved with stealth through an unguarded side door, and waited behind a dumpster until it was safe for the journey to the street. Quincy placed his hand in McKinley's back as if he was escorting him to the courthouse, just in case they were seen. He had parked a Kawasaki motorcycle in an alley for their unrealistic mode of escape, another idea he hoped provided time.

As they stood in the alley, McKinley gave Quincy directions to his bank garage where he had a private car parked. There was no reason to waste time, and McKinley straddled the backseat of the motorcycle, and wrapped his arms around Quincy's waist.

They drove off in a drizzle, taking the back roads that hopefully avoided an expected chase. The drizzle increased, and Quincy agreeably smiled as they whizzed down a blacktop road following directions that McKinley screamed into his ear. At the second stoplight, he spotted the bank, turned right, then left into the underground parking garage.

McKinley pointed to the corner space where a black Lexus was parked. When the banker got off the motorcycle, he noticed the black smudges on Quincy's shirt; it was fingerprint ink from his hands, but he said nothing to Quincy. Besides, the money he'd promised would buy him a closet hung with new shirts. Right now his task was cranking the Lexus. He had a key stored in the heel of his shoe, and leaned against the door, placed a leg over the other knee, and rotated the heel. He and Quincy got into the car and drove off at the posted speed limit.

***

He drove the Lexus to the private airport located adjacent to Kennedy Airport where he had leased storage for his personal light aircraft. A black Lexus dimed off its headlights as it pulled through the gate of a wire-fenced private airport. The air station was unoccupied because the bank had not renewed its contract, but McKinley had the keys to the aircraft. He parked the sedan in a parking space. McKinley reached over and unlocked the glove compartment and retrieved a metal box containing a key. He took out the key and shoved it into his pocket.

The two men rushed out on the paved tarmac, led by McKinley, whose head rotated side-to-side, looking for anything that might interfere with his takeoff. He inserted a key into pilot-side door of a two-seat CC11-160 CarbonCub SS. His body slid into the seat, and he bobbed his head at the man who had granted his freedom. Quincy saluted, pulled back the parking blocks, and back-stepped; he had expected the backwash.

McKinley flipped several switches, checked the aileron movement. He thumped the gas gage, and revived the 180 horsepower engine to idling speed. The modified Cub leaped into the sky at a climbing rate of 2,100 feet per minute, although she could climb to 10,000 feet elevation at a speed of 138 miles per hour. The 25-gallon gasoline tank would take them to Miami, unless he faced a headwind and had to refuel. The aircraft leveled off at 7,500 feet, well below larger aircraft, until he wiggled out of the crowed airspace. He set the compass for the east coast of Florida following the coastline.

When the Cub was airborne, Quincy drove the Lexus off the tarmac, his mind retracing his previous plans. He had completed his task and now he must escape to a hidden hiding place at the Billings Farm, a place not even Alice knew—Alice, just the mental thought of her brought tears to his tired eyes.

***

Henson scratched his head, after he had chewed out the officer in charge of the jail, and went into the Captain's office for his own crucifixion. The Captain sat at his desk reading a report. He slowly raised his head, the wart on his nose wiggled as he frowned. His words were unpleasant.

"A blanket arranged to look like our fugitive—this kind of thing went out with Alcatraz."

Henson bobbed his head. "It's carelessness, simply lack of discipline, sir."

The Captain crunched back in his seat, and reached for his cup of cold coffee. "Discipline is not your problem, Randolph—but you've got to find that murderer."

"Yes sir, I'm on it." His cellphone suddenly buzzed. The Captain nodded for him to take the call. As he listened to the message, his face slightly smiled, and then his expression soured. He closed the cellphone.

"Well, Randolph—let's hear it."

Henson nodded as he sorted what where his options. "That was Peter Meirs, former CIA operative that I've been working with on the Cook case. He says he put the FBI on locating McKinley—they've traced him to Miami."

The Captain brought his fingers together in a profile of thought process. "McKinley has crossed the

257

state border; it's a federal case now," he reasoned, but wasn't too pleased with outside intrusion.

Henson's mind went into calculation mode. "Sir, Peter Meirs, and Dr. Caruso are already heavily involved in this case, and we have no better support than these two veterans."

He separated his hands and dropped his elbows on the desk, caught in checkmate. "All right, it's your baby, just be sure you bring that fugitive back to New York."

Chapter 55

THE DRIZZLE HAD stopped and left random puddles reflecting on the roadway, as Henson motored to Kennedy airport. Peter had mentioned that the flight director in the big tower had recorded an unauthorized light aircraft take off at an adjacent private airfield, without an approved flight plan. If this director had more information, then maybe he had a lead. But he wasn't counting on it–why should he, when he had not asked Caruso to come along. Perhaps he wanted to nail this guy himself. Prissy's evidence placed McKinley in Dr. Cook's walk-in safe. It was strong evidence that said he was the killer of John Cook. Yes, he wanted this man all by himself.

The skyline was lit by the lights of the airport as Henson stopped at the streetlight of the entrance to the airport lobby. The light finally changed green and he turned into a four-lane entranceway, and motored to the boarding flight curb. He parked and slipped out of his seat. An attendant dressed in a long overcoat that glisten wet, met him on the sidewalk.

"You can't park here, sir."

Henson flashed his badge. "I need to speak with the flight director right away."

The attendant focused on the badge; it was not unusual, only the request. "Just one moment, sir," he said, and raised his cellphone to an ear, and spoke three short sentences. He lowered the cellphone. "All right, park your car in the VIP slot, here's a tag for your departure."

Henson stretched out his hand and gripped his aging fingers. "Thanks buddy, when you're downtown, stop by Midtown, and we'll have a snort together."

A wide smiled sliced across his beard-stumbled face. "Yes sir, yes sir, indeed," he replied with a tip of his hat. "Thank you, detective Henson."

Henson felt good inside for a change, the tension had relaxed its grip on his tense muscles, and even the headaches had disappeared. He had learned so much from Dr. Caruso; that man made friends wherever he went, and maybe his charm had rubbed off on him.

***

A man opened a door down a hallway, and motioned to Henson who sat in a row of chairs waiting to see the flight Manager. He stood and strolled to the open door. The man guided him to an office with opaque glass. Henson opened the half-glass door and faced a secretary seated at a desk filing her nails. She was pert, jovial, and had a disarming smile.

"Mr. Anderson is expecting you, detective Henson. Walk through that door at the end of the hall," she smiled, pointing a nail clipper without taking her eyes from her nails.

"Thank you very much."

She nodded her almond-shaped head in approval, casually filling her nails.

Henson trudged down the hallway, and swung open the door. His face wrinkled with surprise. "Why Peter!"

"Sit down, Henson. Mr. Anderson has something to tell you," Peter Meirs said.

Henson was familiar with Peter's surprises, but somehow this one was out of character. He settled in a

260

leather soft-cushion chair seated in front to the desk. Anderson rounded his desk and sat in and adjacent chair, rocked forward on the elbows of his checkered jacket.

Anderson unscrewed the cap from a bottle of water. "Yesterday a CarbonCub SS took off without a flight plan, your man, McKinley, was in the pilot's seat according to ownership records."

Henson frowned; yet, his face expressed calmness. "We knew McKinley had a light aircraft, and were not surprised that he took flight, only the method of his escape—where is he going, that's what I want to know."

Anderson swallowed some water. "We may be able to help you on that, too, but there is a more urgent matter. Mr. Meirs, why don't you bring Henson up to date?"

Meirs close the cover on a Zippo lighter embossed with the Marine insignia, and blew a few smoke rings, his face unmoving as if its facial muscles were glued.

"We had hoped that Mike Quincy was with McKinley but we can't confirm it at this time.

The drama intensified.

"The latest information says McKinley was alone. It looks like we have a loose end; Quincy aided McKinley's escape."

Henson's emotions exploded, and he stood abruptly, pounded his fist on the desktop. "This McKinley guy is a cool character, and must be apprehended!" His thoughts convicted his actions that had set Quincy free from the safe house. These misfit cousins and close relatives are one for the book, he thought.

Anderson stood, loosened his collar. "Detective, your man took off this morning from Miami on a direct airline flight to Peru, South America. Mr. Meirs has had the FBI on his trail since the flight left Miami."

Meirs crushed the cigarette butt with a twist in an ashtray. "If you would like to come along, we can extradite McKinley back to New York."

Henson swallowed his anger—it was exactly what he must do to satisfy his Captain. Meirs was Dr. Caruso's trusted friend, and that persuaded him.

"Listen, Meirs. I don't know how to repay you and Caruso, heaven knows I want to. But it's time that I take control of this case, or the Captain is going to chew my butt."

Meirs nodded. "Henson, I wondered when this moment would come. Nobody likes to be upstaged."

For the first time, he felt professional about the case. Not since he was twelve, being with his father on the fishing boat with a hook stuck through his thumb, which his father removed, had he realized such satisfaction. Henson bowed his head, shielding his embarrassment. It wasn't a case of upstaging, but a case of two men he had learned to trust and respect. Dr. Caruso and Peter Meirs were uncommon men; each of them, in some way, reminded him of his father.

Chapter 56

STANDING ON THE flight deck, Henson closed his cellphone after reporting to the Captain. Fortunately he was forgiven for the escape of McKinley, especially when he learned that McKinley would be extradited to New York. The attendant who arranged the VIP parking placed walked up and greeted Henson with a smile.

"Well detective, leaving us so soon?"

"No, I'm thinking about boarding a direct flight out of the country."

The attendant's round belly shook as a chuckle roared into a guffaw. "It won't solve anything," he said suddenly laughing.

Henson chuckled, releasing the guilt of his embarrassing outburst in the flight director's office previously, and the speech he'd made to Meirs. This insane case would cancel his promotion, if McKinley escaped. He felt the tension, the pressure in his chest, and remembered the sleepless nights, the long days without palatable food. This attendant thought he was simply escaping the tension of his life. What a joke.

"Guess you're right philosophically speaking. Just have to stick it out here in America I suppose," Henson smiled with a slight grin. A thought suddenly hit him square in the eyes. Even his actions mimicked Caruso.

The attendant chuckled. "This is a great land, detective; it's a people problem, not America–we've strayed too far from the founding fathers beliefs."

Henson mused at the remark until his peripheral vision saw the flight director waving in the tower. "Don't forget–in my office for a snort."

"Right on, sir."

The flight director reported a direct flight to Peru in about an hour. Henson's appetite decided he'd head to the café in the lobby for some food. And then it struck him in the stomach, he may upchuck after the pizza he had had with Peter.

***

The flight finally arrived, and Henson went to the lobby newsstand for a newspaper. He tucked the paper under his arm as he rushed down the corridor to the gate of the flight. When he arrived, a line of waiting passengers had already spilled into the walkway of the Delta gate. He stood at the rear of the line, and surveyed the headlines of the newspaper.

There was no report about McKinley's escape. Peter had frozen the story to enhance McKinley's confidence. Confidence gave the opportunity for a mistake, proper police procedure.

Henson inched forward in the line as he reviewed McKinley's words at the interrogation; weasel words that had destroyed the force of law, and now he had taken the law into his hands. Yes, McKinley was a weasel that ruins an egg by sucking out its contents leaving the shell intact. Many lives were simply empty shells of ruination left by vaulting greed. The attendant is correct, he thought. America is suffering from greed and arrogance.

Suddenly Chip Cook rounded the corner, an overnight case swung from his hand. Henson scratched his head wondering who the culprit was that had called Chip. Perhaps it was Peter, perhaps

Caruso.   Yet Chip had the strongest motive for recapturing McKinley.

Henson smiled. "Well, Chip. Are we still trying to bait a hook?"

Chip slapped Henson on the back. "No. He has slipped through the net, but this time we will reel in this fish for good."

The lingering line of passengers finally pushed forward, and the two men entered the covered canopy leading into a Boeing 737 jetliner.   Henson took the seat next to the window, and Chip nestled in the aisle seat and donned the earphones; Henson locked his seatbelt anticipating a turbulent flight, not the airwaves, but the chase of McKinley.

The seatbelt light lit with a dinging sound.   An airline attendant waltzed down the aisle and assured that each seat was in up-position with seatbelts locked. Chip removed the earphones and tightened his seatbelt around his thin waist.

Suddenly the pull on the belt at takeoff seemingly created a noise that stirred in the shadows of his distant vision.   The noise came from the four-wheel drive he and his father had taken into the wilds of Tanzania on that last trip to the foot of Kilimanjaro Mountain: The beauty of the barren land, the open sky, nights under the stars, and the howl of distant hyenas silhouetted in the light of the moon.

His eyes suddenly popped open.

An attendant was lowering the tray on the seat back. "Can I get you something to drink, sir?"

"Why . . . why yes, coffee, black," he stuttered.

Henson raised his head. "I'll take coffee, too, one cream."

She nodded with a bright smile.   "Thank you gentlemen, enjoy your flight."

Chip thoughts drifted to McKinley.   The information from Rosie stuck in his mind, a moment to review everything she had told him. But surprisingly he wondered whether the man had any kids or relatives. He had forgotten to ask, and somehow it was important—just why, he had no answer, only a hint of compassion had surfaced from the depths of his teenage years. He flipped open his cellphone and hit a preset button. The phone rang at Mary's desk.

"Cook Medical Center: How my I direct your call?"

"Mary, its Chip."

"Chip, heard you were on a flight to Miami!"

"Sitting right here on an aisle seat—is Rosie available?"

"You just caught her. Hold please."

Chip gazed out the window until the phone clicked, a quandary of thoughts rushing through his mind on one subject: McKinley: who is he really?

"Hello, Chip. What's on your mind?"

"Thanks for answering my call, Rosie: one question. Did McKinley have a family or relatives?"

"One brother who was killed in Vietnam—he never married."

"Really? He must have lived a lonely life."

"Yes, I think your father was his only friend."

"Thank you, Rosie—I'll call you when we get back."

"Have a safe flight, Chip."

***

Dr. Caruso had heard of McKinley's escape and Quincy's assistance. The situation churned in his analytical mind and he reasoned there were still a few loose ends in the case. He called Prissy, and she had agreed to meet him at her laboratory in Rochester. His reasoning had settled on Quincy. He had been let off too easy in Caruso's opinion. There were more questions in his mind that needed answers.

As he drove through the northern countryside, he thought of Chip's trip to Peru, and was consoled that Henson was with him. Meirs call told him that Henson was his own man. And if his hunch was correct, there might be great confusion in South America, a diplomatic event according to his call from Cavits in Washington. The Peruvian government had offered it's assistance with the drug war in Bogotá and the President feared an embarrassing scene might disrupt the diplomacy. It was Peter's job to circumvent that disruption. In his mind, he favored Henson's idea. The purpose of diplomacy had always been to prolong a crisis. Henson would operate on the margins of diplomacy, and still prolong the crisis.

***

Caruso finally arrived at the Rochester forensics laboratory and found Prissy munching on a sandwich in the lobby, her packed handbag sat beside her. She smiled as he approached.

"What's this hunch of yours?" she beamed.

"Don't quite know myself, but I think your assistance will be helpful in finding the answer."

"All right, let's be off," she replied and gripped her handbag.

They motored through the rolling hills and finally arrived at the Billings Farm after stopping only once for gas and a meal. As they drove into the driveway they noticed that the front door was open. Caruso quickly parked and rushed to the door.

Obviously someone with a key had opened the door; perhaps, the police, yet they weren't as clumsy as this, he reasoned. They entered the office and went down a hallway to the house attached to the back of the store. Prissy went into the family room, Caruso ventured into the master bedroom. His keen eyes looked for evidence locked in his mind, evidence so scant, it was thin and elusive. Suddenly he spotted what looked like a diary, lying on the mantle above a quant fireplace. Curiosity gripped his thoughts as he called Prissy.

She entered the bedroom with her handbag, and Caruso asked for a pair of vinyl gloves. She immediately reached into the bag and pulled out two pair of gloves. Quickly they donned the gloves, and Caruso fingered the diary. He laid the book open on the bed, and fanned the pages, Prissy's eyes excitingly staring over his shoulder at the pages.

Suddenly the pages opened to the last entry. Both pair of eyes glued to the passage. According to the notation, Brian Latham had left a package in his sister's keeping. Alice had written this passage and recorded that the package was in her nightstand by the bed. Prissy quickly strolled to the nightstand, and slid out the drawer.

There on the bottom of the drawer lay a case about the size of a package of cigarettes. She turned to Caruso, snapped a picture with her camera, and lifted the case from its hiding place. With great expectancy, Caruso snapped open the case, the lid swung out on a tiny hinge.

Both pair of eyes gapped wide open.

A 10-cc syringe nestled in a felt slot beside a vial fixed with an injection septum that lay beside it. Caruso gently took the vial from its slot. He raised it to his eyes, startled by the contents: Sarin!

Caruso raised his cellphone and punched a coded button, while Prissy snapped pictures for admissible evidence. The tiny speaker clicked through a series of relays to overseas satellites. Finally, a voice answered emanating aboard a Delta flight high above the Pacific.

"This is Henson, Dr. Caruso," he said, familiar with his cellphone number.

"Listen carefully. McKinley may be innocent. Don't allow the press to get involved, whatever the circumstances. There may be a diplomatic problem."

"Understood." Yes, he understood but did not comprehend. What new evidence did he have? If McKinley is not John Cook's murderer, then who is?

Chapter 57

A DELTA AIRLINER lowered its flaps over Dubai International Airport, landing lights piercing through the darkness on a ruler-like airstrip bordered by a parallel row of lights. Passengers inside the airliner, mostly Peruvian tourists on vacation, prepared their seats for landing as attendants stood at the exit door. The squeal of the wheels scorched the pavement as the gliding aircraft revved it turbines in the standard practice if an emergency required aborting the touchdown. The airframe settled on the runway like an eagle with widespread wings settling on its nest, and reversed its turbines thrust.

The cumbersome airframe crossed taxi lanes in its zigzag trip to the tarmac under the power of idling turbines. Finally, she braked and swung toward the airport structure guided by the moving hands of an air-traffic attendant. Easing toward the wall, the attendant dropped his hands, and the pilot immediately applied the hydraulic brake, and cut the turbines. The noise of rattling luggage trailers as they whizzed under the rear door, and took on its cargo. A bellows extended on a telescoping track and bumped against the fuselage door as an airport attendant ran down the covered canopy and locked the bellows to the fuselage. The door of the aircraft swung open, and passengers streamed toward the lobby, each watching for the signs directing them to the luggage carousels.

McKinley left the sprinting passengers, and entered a corridor to rent a car. A barefoot young lad dressed in short pants frayed around the edges snatched his receipt, and offered to get his luggage. McKinley cocked his head peering into the boy's eyes. A stored fantasy of when he was a boy flashed for a millisecond in the recesses of his mind. Why not, he thought and gave the lad ten American dollars.

He pondered the wisdom of his decision as the lad ran off to the carousels. Although he had never married, he always wanted a son; it was his secret fantasy, another reason he had admired John Cook and his son. The thought merged into the memory of how and why he had lost his friendship with John. Dyspepsia stuck in his throat. It hurt. Was it remorse or jealousy? He turned, and hurried through the corridor to the rental counter.

The young lad met him at the rental entrance of the corridor with his luggage, and McKinley gave him another five dollars for information to Cusco, the sacred valley of the Incas. The lad responded with a brochure he took from the counter.

"Me show way, Senior," the lad responded in broken English.

He was about twelve, dark skin with deep brown eyes and black unkempt hair. The lad apparently had no parents and made his living fleecing American tourists, but he had been honest enough to bring his luggage and had not run off with it and the cash.

"Yes, how long will it take to motor to this valley," McKinley asked?

"Pardon sir, better you take autobus—leave every day to hotel in de valley."

McKinley smiled. "What's your name, son?"

He thumped his fist in his chest. "Me Mario, good guide."

McKinley placed his hand on the lad's shoulder. "You are hired Mario."

The lad's charming face beamed and he took his client by the hand. He led him out of the airport to the curb. He left McKinley standing at the curb, while he ran off along a brick surface, snow-capped mountains rising in the misty distance.

As McKinley waited, he sat on his luggage, surveying the milling people, and comparing them with Mario. He decided they were mostly bighearted illiterates who invented dreams, and took refuge in illusions because few actually read books. And he wondered how Mario had learned to speak English. Most of the languages were Spanish, Aymara, mixed Amazon dialects, and Quechua, the official language of Peru. Oh yes, he suddenly thought—he learned English from the gringos like himself. Mario was an industrious street urchin who made his living on naïve tourists. McKinley was not naïve, but this young lad had his attention.

McKinley's mind seemed captivated by the primitive beauty of the Peruvian terrain, its floating lakes in the highlands, glaciated Andean peaks where condor's soared, cocoa shrubs planted down its cascading slopes, and in the plateau valleys populated by tropical birds. What a paradox existed in the cocoa plant: the source of hot chocolate from the beans, and cocaine from the leaves. Yet deep in the matrix of McKinley's mind reality suddenly stormed to the forefront of his thoughts.

He would disappear in some off-beaten path in the middle of nowhere, a place where he could collect his thoughts in safety. Questions needed answers: How would he circle back to the city and take a flight to Switzerland? Then a stumbling block hit his mine:

272

How would he escape if the American authorities found him before he implemented his plan?

Mario stood about thirty feet away in the shadows waving his hands in a desperate attempt that finally wrestled McKinley's attention, lost in distant thought. McKinley shook his head; the cobwebs cleared, and he stood, collected his luggage, and strode over to Mario. The young lad took McKinley's hand, and pointed at an autobus parked beyond a row of columns.

"That your ride to Cusco."

McKinley nodded, and shook Mario's hand. "Take this Mario, and thanks for your help," he said as he extended a twenty-dollar bill. His friendly smile was indelible, and a subtle thought crossed his mind: Could Mario help him escape?

Mario waved as McKinley boarded the bus. Somehow the little lad wondered if this man perhaps needed his assistance before the next dawning of the Inca sun.

***

Loaded with eager tourists, the autobus pulled up to a hotel, a modern building set in the heart of the sacred valley of the Incas, once the capital of the extensive Inca Empire. Picturesque mud huts were staged all around the sacred grounds set in labyrinthine caves. Ancient ruins seemed to expose their prehistoric secrets.

As McKinley sat in his seat, exiting tourist casually poured past his window. His mind soaked up the historic view beautifully staged in a panorama scene, hidden in deep shadows cast by the long rays of a late afternoon sun. And just to think that these ancient people had built an empire despite the absence of the wheel and a system of writing, and still achieved a high-level of civilization. McKinley had consoled his desperate thoughts by reading a brochure left on each seat. He took the brochure and stepped off the bus, and found himself staring at a captivating landscape.

The Incas showed skill in terracing, irrigation, and mining, and produced fine textiles, metalwork, and pottery. They were governed by an absolute monarchy, supported by a ritualized hierarchic religion. The Inca Empire was engaged in civil war in 1525-32 AD, when the ruling government was weakened, and even the food supply had dwindled. By the time of the invading Spaniards under Pizarro, there were few defenders. Even though some Incas revolted against the invaders, the Spanish flag flew over the empire until the 19th century.

McKinley stood in the rather long line at the hotel desk as his mind cleared, and realized he must finalize a plan. First he'd register under a false name, and dress in native clothes, and then he'd trek out into the jungle and wait until it was clear. With this overall plan stirring in his mind, the line moved until he finally stood at the registration desk. He paid cash for a week's rent, since he planned to use the time to plan a path that led detective Henson and Dr. Caruso away from his location. Then he'd circle around; make his way back to the airport. Then a snag in his plan hit him. He had no transportation. He sighed deeply, oh well, he thought, he'd think about that later; these people seemed to walk everywhere, perhaps he'd walk, too. He took his room key and ambled to the stairway.

McKinley opened his door, tossed his luggage and briefcase on a chair, and crashed exhausted on the bed. There in the quietness, his mind automatically shut down his body. He required sleep, but sleep would not release the remorse he now felt about how her had treated John Cook; of course, Quincy was helpful and he had been paid handsomely. His lungs released a deep sigh, and his eyes fluttered irresistibly closed.

He somehow floated in midair, visualizing an airliner circling in the clouds over South America— Henson must be near, replied a soft voice deep in his subconscious. But a worse problem loomed in the jungles, where his body floated near sacred Inca relicts. Yet, he floated deeper into the greenish shadows, drawn by the ancient call of fierce Inca warriors.

Suddenly the awareness of idol gods hidden in the deep jungle gripped him in a stark nightmare. Ancient prehistoric flora divulged their secrets as long vinery fingers curled over his head in attack formation. Huge snakes and eerie vermin slithered in the grass, their fangs exposed in grotesque expectancy. The fangs struck, and pierced though his flesh!

His eyes fluttered opened!

Weird thoughts clogged his mind. He rolled over into another position, too sleepy to worry. Dawn was only a few hours away, and he needed every ounce of energy if he hoped to escape the manhunt he was certain that would greet him very soon.

Chapter 58

A NOISE OUTSIDE the hotel room awakened McKinley, not the nightmare that tortured him all night, an animal sound real and alive. He rose from his bed, and walked to the window and looked outside, but all he heard were tropical birds singing their mating calls. Perhaps a bird in the tree near his window had made the noise. A close examination revealed that he was correct, a brilliantly hued tropical bird sat on a nest woven in a forked branch; the other noisemaker was a cat on the ground that had been rudely peaked on the head.

A smile crossed his face as he turned from the window and glanced at his wristwatch. The time suddenly startled him until he realized he had not reset his watch for the time-change between New York and Peru. Silly of him, he thought as he twisted the dials. The time now was nearly seven o'clock in the morning. He dryly swallowed and his stomach announced his hunger. He dressed, threw a few items in his briefcase, locked the door, and went down to the cafeteria.

Fortunately the cafeteria was opened, and he sat near a window and opened a menu on the table. Three languages were printed by each item, Spanish, Quechua, and English. As he read the menu for a few seconds a young waitress suddenly appeared.

"Buenos Dias."

McKinley nodded, not sure how to respond, and he pointed to the menu, and apparently ordered eggs, toast, and coffee. He had no time to waste, and as she poured his coffee, he glanced over at the brochure he'd taken on the bus. It gave directions to the sacred valley, and Machu Picchu where he planned a temporary base until he assured the coast was clear for his escape. After he finished his breakfast, he ordered a sandwich and stuffed it in his briefcase.

**

A lone hiker marched down a dirt trail as the brilliant sun filtered through the leaves and lit his pathway. In one hand he held a briefcase by its handle, in the other he gripped a long stick, not a weapon, but a support for balance and perhaps to ward off snakes. He followed a map printed in the brochure.

As he trudged along under a steamy canopy of rainforest greenery, he finally reached the pinnacle of a rising hill. Suddenly McKinley's eyes were enthralled by the snow-crowned Andes Mountains spread upon the horizon like a continent displayed in miniature. The distant slopes seemed scarred with deep gashes from the weathering of annual flows of melted snow. The majestic wings of condors soared above its glaciated peaks.

The prehistoric view continued to seize McKinley's awareness; simply breathtaking, alluring calls from the deep, dark rainforests of the Amazon basin. As he topped the hill, he took giant steps down the rocky slope on the opposite side until he reached the valley floor. Immediately he was overcome by the size of a giant volcano rising in the distant cloudless sky. Beyond the valley he heard the crashing waves on an unseen beach, and heard the echo of argumentative sea lions barking as they lavished on the sun-glazed outcroppings of the Pacific. Suddenly he fixed his eyes on the target of his trek into the wilderness. The sacred valley wrapped around the basin like a giant meadow of sand, the Machu Picchu ruins bathing in the torrid sun. He stumbled into the desert valley between two giant sand dunes, and collapsed against an ancient hewed rock at the fringe of the rainforest.

Before he caught his breath he noticed the movement of dried leaves in the crevice of the rock. Unaware of the reptiles in this lower Amazon valley, he cringed with fear. Sudden fright bloomed to unabated terror as he snatched his hand from the sandy ground, his wide-open eyes focused on two fang marks. The triangular head of a Bothrops viper slithered away under the vines. Sudden pain penetrated the wound as he squeezed his wrist with the other hand. Dizziness defocused his blurred eyes, and he fainted.

***

The sun rose and set twice, and McKinley finally regained consciousness. The pain pulsated in his grossly swollen hand, and he found the he had wet his pants. Then he wondered how long he'd lay unconscious. He remembered reading the brochure about snakebite and thought some varieties of snakes affected the kidneys, but he hadn't realized that he had only a few hours left to live, and then agonizing death.

An eerie sound in the rainforest rang in his ears, but he could not move. He lay in torment while his frazzled mind danced in a kaleidoscope of lights, no sense of time, lost in uncertain limbo. Deep in his subconscious mind events that he'd stored for years danced in the blackness of nowhere; the friendship with John Cook, his dates with Jenny Lynn before they were married. How John's genius had expanded a small clinic into several buildings with five physicians and six nurses, one his wife. As the practice grew, his bank had financed the Medical Center complex. Finally McKinley broke off all relationships. Yet the anger persisted, but remorse festered in his mind and drove him into a neurosis that had required two years of treatment, which he had never told anyone.

A shadow burst on the scene.

A small boy knelt and laid the snake bitten man's head in his lap, stroking his hair as he wiped the sweat from his brow with his pullover shirt.

"O gringo, you wait not for me," Mario whispered, tears cascading down his dark face.

McKinley's eyes fluttered open from a deep abyss, a limp arm fell on the boy's shoulder. His stubbiness challenged the respiratory paralysis that had invaded his lungs, and the neurotoxin that seized his speech as he tried desperately to open his parched lips. Words finally reached his tongue.

"Mario, you are good kid . . . in my briefcase . . . fifty thousand dollars—you take . . . get education."

The limp arm slid from Mario's shoulder. He lost consciousness.

Death lingered.

Mario leaped from the ground, rested McKinley's head on the rock and dashed into the rainforest as he wiped the tears from his cheeks. His urgent task was getting an anti-venom serum. He knew the hotel had a resident physician because he'd brought him supplies on several occasions.

He ran as fast as his legs would take him, until he found the shortest trail back to the hotel. All his thoughts were focused on how much time the gringo had to live, and approaching darkness was sure to complicate rescue. The money never crossed his mind, only the life of this gringo. And yet, he'd only met him a few days ago. He knew nothing of his past, only that he was an American in a strange land, quite obviously he knew nothing about the dangers in the Amazon rainforest. But in his eyes this gringo was a good man, and he had to do all he could to save his life.

As he galloped into a sudden clearing he saw the hotel, and raced to the rear entrance. The light shone through the physician's window, but he saw no one. He burst through the door, his keen eyes searching the room.

The physician was not at his desk or in the back room.

Mario rushed to the cabinet where the physician kept the serums for snakebite, some refrigerated. Seven different anti-toxin bottles were lined in a row. Fortunately Mario knew the snakes in the area and selected the bottle he thought would keep the gringo alive, until he brought him to the physician. He grabbed the bottle and thrust it in his pocket with a hypodermic needle, turned abruptly and ran out of the office, the door swung wide open behind his exit. Two men and the physician approached the door as Mario ran out. It only took a few seconds to relay the emergency as Mario's hands gyrated wildly.

"How long has he been unconscious?" the physician asked.

"Only 'bout three hours but me don't know how long he snakebite when me arrived." Mario stuttered.

The physician dropped his gaze on Mario, and stepped into his office and grabbed his medical bag. He told the two men to get a stretcher and follow Mario. They entered the rainforest darkness, led by Mario and a battery lamp. As they ventured through the heavy vines for a good half an hour, Mario finally led them to a quicker trail than the regular path to the sacred Inca ruins. He knew time increasingly ran out for his gringo friend.

A mountain cat suddenly crossed the path, and Mario quickly thrust the beam of his light in its eyes. The cat jumped completely over the path into the cover of the vines, and disappeared in the eerie shadows. The eyes of the man with the stretcher were wide open as he gasped. The physician smiled.

Mario finally saw the ruins, and guided the men to the place where the gringo lay. As they approached the unconscious man, the physician quickly removed a serum bottle and a syringe from his case. Mario lifted the gringo's head as the physician rolled up a sleeve. He stuck the needle of the syringe into the rubber septum and drew out six milliliter's of serum. Mario had already taken out a swab of cotton soaked with alcohol, and gave it to the physician. He quickly rubbed a patch of skin over a vein, and thrust the needle into the blood stream.

The physician breathed a deep sigh, and placed the items back into his case. He instructed Mario and his helper, and they placed the gringo on the stretcher as he explained the urgency. Since he didn't know the species of the snake, the serum may not be effective enough before his kidneys erupted. The helper laid the stretcher on the ground beside the gringo, and they rolled him on the stretcher, and then positioned him on his back. Mario spread a blanket over him to ease the chill. The helper and physician lifted the stretcher as Mario took the medical case. He ran out into the lead and placed the beam of his light in the pathway. As they rushed toward the hotel, Mario wondered what he'd do about the briefcase the gringo had insisted he take. As his feet pounded the rainforest floor, he pondered the responsible action. It was more money than they had ever seen, surely enough for him to attend University. But was it the moral thing to do? He thought of the priest at the catholic church how had talked with him on many occasions. What would the priest do?

The physician stopped and signaled rest for a moment, while he checked the condition of his patient.

Mario and the helper sat on a decaying log as the physician took the pulse and temperature of the man. Beads of sweat rolled off the forehead off the gringo. His temperature was 103 degrees Fahrenheit, his pulse rapid. These signs indicated the man needed more serum, and hopefully his office had a more effective antidote.

They finally emerged from the rainforest and went directly to the physician's office. Mario knew the gringo was in good hands, and went to the hotel; he just couldn't watch the gringo die. As he approached the front entrance, he noticed an official car parked in front. And through the glass door he saw an *oficina de policia* standing at the desk with three gringos. He immediately thought they were looking for his friend with the snakebite. Questions flooded his mind. Was the gringo a fugitive, had he stolen the money?

Mario entered the front door and strode to the kitchen, his ears perked to the conversation at the desk. From what he gathered as he passed, the three gringos were looking for someone. And he suspected it was his friend.

Mario left the hotel and went back into the bush. Finally he found the gringo's briefcase, and quickly returned to the hotel. Mario cleared away a few glasses, and laid the case on the table. He dared open it, but urgency decided that he must. He stood for a long moment with agonizing thoughts: this money would provide his pauper's livelihood for many years. Vivid pictures flashed through his young mind as he finally opened the case.

His beady brown eyes gazed at one of several bundles of $100 bills in American money. In a sudden moment his conscience gripped his thoughts.

It was wrong if he took it.

Mario closed the briefcase, gripped the handles, and stepped through the kitchen door into the hotel lobby. The three American gringos were still conversing with the *policia.* He deeply sighed and walked toward the desk. As he approached, the registrar turned to the *policia.*

The clerk gazed at the lad walking toward him. "This is Mario. He can answer your questions."

The clerk stooped to Mario's height. "Mario, these three gentlemen are from America. They are looking for the man you met at the airport. Can you help them?"

Mario bowed his head. "Can do," he replied and raised his head. "Will gentlemen follow Mario?"

The men followed Mario's lead out of the lobby. As they walked along in the warm breeze, Peter whispered into Chip's ear. The words were electrifying if McKinley was innocent! Peter tugged at Henson's shoulder and mumbled a few words. Henson's eyes showed disbelief, the case still far from resolved.

Mario led the men to the rear of the hotel and pointed to the physician's door. "The man you seek in physician's office."

Mario opened the door for their entrance. The physician had heard Mario's voice, and met them at the door. Introductions were brief, and Chip Cook described the reason of their visit. The physician glanced at Mario and nodded. "I'm afraid you are too late. He died of snakebite a few minutes ago."

Shocks seized awareness.

Sadness wrinkled Chip's forehead, the expression of injustice. Henson strolled to the bed where laid the deceased. It was Morris McKinley all right. And so the search had ended in Peru. He turned to Peter. "It's him all right," he sighed.

The detective sighed deeply, the expression of a finished case, and he addressed the physician. "Would you arrange for the transportation of the body to the

283

airport, and airship it to my office? Here's my card," he said extending a business card.

"Certainly," the physician replied, "along with an autopsy report, too."

As the men left the office and went back into the hotel, Mario purposely followed behind. And when they reached the desk, Mario tugged the hem of Chip's jacket.

"Pardon, sir . . . Mario speak with gringo, okay?"

Chip turned, and stooped. He noticed the briefcase dangling in the lad's hand, yet he saw the expression on this young boy's charming face; studious eyes, a strong countenance, sturdy frame, abundant confidence.

Instantly Chip's mind flashed to the moment he had walked alongside his tall father at Mount of Kilimanjaro. Just a lad filled with excitement and wonder, until his father was murdered. And here in this distant land he had chased a man he thought was his father's killer, and still the evidence was inconclusive. And yet somehow his entire cognizance focused on this young lad, who knew more about McKinley than he. Somehow he was relieved that McKinley was apparently innocent. His hidden compassion resurfaced.

"What can I do for you, son?"

Mario handed over the briefcase. "Gringo, he give me dis briefcase. He say me keep it, but me think I not. You take please?"

Chip's face registered puzzlement: the honesty, seeming maturity of this young lad. He opened the case, and saw the large sum of money, and then gazed into the lad's face.

"What did McKinley say, Mario?"

"He say me take for education."

Chip looked into the lad's eyes for a moment, closed the briefcase, took Mario's hand, and stood as he turned to the counter.

"Does Mario have parents," he asked the registrar.

"No, Mario lives here in a room next to the physician's office," he responded, yet he decided to reveal more, "he makes his money at the airport helping gringo's like Mr. McKinley."

Chip's face blushed; his mind suddenly registered a quote from his father during the last Mount Kilimanjaro trip. The words rang in his ears, *"Someday, son you will make a fine physician."*

Chip realized at that moment no physician had a more rewarding opportunity than to assist this young lad who stood before him. He stooped and placed his hands on Mario shoulders.

"Mario, I make you a proposition."

His eyes glazed. "You give . . . how you say . . . opportunity?"

A slight smile creased Chip's cheek. "That's exactly the word—how would you like to come to America. This money will go into a trust fund for your education."

The boy stood completely overcome by the generosity of this gringo, though Mario's mind was not on his education. From his brief experience, he thought this Mr. McKinley was a good man. Were all American gringos like this man?

"You take Mario to America, Si?"

Chip nodded. The joy in the boy's face, the smile marred by dirty cheeks; this was the look of humble thanksgiving, the gift of one man to a deserving child. And the truth was that McKinley had been a good friend of this boy, a good friend to his father also; yet, greed stole the friendship and drove McKinley to the moment he remorsefully confessed to Mario.

Somehow Chip understood. McKinley had trusted Mario, and that trust had led to friendship. But now he understood that McKinley's confession was for him, not Mario. He harbored no grief, and hoped that McKinley had understood. But he made a decision then in the presence of this young lad, who reminded him in his youth. I needed father, but he was too busy and often away from home. Chip had decided what to do, and he would complete McKinley's wish. The money belonged to the bank, and McKinley was an officer of the bank, and he'd paid for his crime with his death. It made perfect sense to him.

Yes, Mario. I will send for you the moment I reach America," he replied as he handed a business card to the registrar. They briefly discussed the procedures for Mario's trip to America: Chip would contact the American ambassador, the registrar agreed to contact the Peruvian Magistrate.

Henson had said not a word or even challenged Chip's decision. He wrestled with a seemingly unsolvable case. His thoughts raged to fever pitch when he reasoned what evidence Caruso had discovered that exonerated McKinley. Yet, he knew that Caruso had good reason for his actions, and challenged himself to wait for the evidence. Besides, McKinley could no longer defend himself. Caruso must have uncovered evidence that pointed to the real killer.

Chapter 59

PRISSY ENTERED HER forensic laboratory in Rochester, and immediately sat the gadget bag on the bench beside her computer. She quickly donned a pair of latex-free vinyl gloves and carefully took a metal syringe case from the bag, and laid it on a sterile towel. From an overhead shelf, she took a role of special tape, and stretched off a section, and then stuck it over an imprint on the case. Prissy gingerly pealed it off the surface, and gazed at a perfect fingerprint.

Facing her computer she opened a file containing the two fingerprints she had taken from the Cook safe in the basement laboratory, and placed them adjacent to the fingerprint taken from the cup used by Morris McKinley during interrogation. Prissy punched a key, and the three fingerprint images appeared on a big screen installed on the wall. She focused the image, and typed several keys on the keyboard.

A program began a search, and suddenly flashed with a 10-point match on two of the fingerprints. A photograph swept across the screen: Mike Quincy on the syringe case and the Cook safe!

Prissy flipped open her cellphone and punched a coded key. The circuit buzzed, and a voice spoke.

"Caruso here."

"This is Prissy, Dr. Caruso. You are not going to believe this."

"Hit me with it Prissy, Henson is on my back," he chuckled.

"Two fingerprints belong to Mike Quincy and match the two prints in Cook's safe and the syringe case."

"Bingo! That's hard evidence, Prissy–good work! Email a copy of the report to my cellphone."

"Can do."

***

Caruso called ahead to the NYPD Midtown Precinct, and found that Henson had arrived in New York from the trip to Peru and was in route to his office after landing at Kennedy Airport. They expected him back in his office within thirty minutes. Caruso quickly engaged a taxi to drive him to the downtown precinct. As they drove along, he mulled the information over in his mind. It was a classic case of circumstantial evidence wrongly pointing to a suspect. It was his error that suggested Morris McKinley as the suspect, but hard evidence disproved it.

When he entered his office, the detective sat stoic at his desk, a cold cup of coffee beside his hand with his fingers drumming rhythmically on the glass top. Henson heard the sound of footsteps enter his office, and turned his head.

"Well, Dr. Caruso, I hope you have some useful evidence. You did not return my call."

Caruso smiled. "Let me get you a hot cup, and I'll tell you a wonderful story," he replied as he took the refill from his desk and poured a teaspoon of his cold coffee into the trashcan. He saw the coffee pot when he had entered the office, and he glided over, and poured two cups.

He sat in a chair opposite the detective and placed his cup on the desk beside his hand. As he sipped his cup Caruso stared into Henson's tired distrustful eyes. He imagined the questions stirring in the detective's mind, the frown on his face, when he had received the news of McKinley's innocence. He sat his cup on the edge of the desk; unnecessary suffering was not in his vocabulary, but it was tasteful, because he had the panacea.

"Sorry that I didn't return your call, but all the evidence was not complete yet."
Henson defiantly crossed his arms over his chest. "This evidence had better be good."

"Better than even I expected, Henson. Prissy and I went out to the Billings Farm on my hunch. We found the evidence we've been searching all along hidden in the night table of Alice Quincy's bedroom," he announced and tossed the diary on the desk.

Henson unknowingly released his cup as he opened the diary, his stomach churning. He fanned the pages to the last entry. As he completed reading the notation, his head raised with a guarded smile. "If Brian Latham gave Alice this case, doesn't that implicate Brian in Cook's murder?"

Caruso rocked forward in his chair and pointed to the entry. "Notice the difference in the handwriting?"

Henson's head lowered, his eyes scanning the entries. His forehead wrinkled and his hand gripped his unshaven chin.

"They certainly look different, don't they?"

"According to a handwriting expert, who compared the handwriting on some farm documents with this last entry, the handwriting is that of Mike Quincy—he's unwarily signed his own death warrant."

"Quincy! Who would have thought it? And then it hit like a ton of bricks. "We've got to find him before he leaves the country," he screamed as Caruso, flipping open his cellphone.

"That's the Henson I knew," Caruso smiled, suddenly laughing.

Henson punched a code, and a distant cellphone rang. "Peter, listen carefully, Mike Quincy is our man, and he may be at Kennedy Airport . . . right . . . okay! Caruso and I will head out there. Thanks Peter."

"What is it," Caruso inquired, releasing his hands from the case. Henson closed the plastic nuisance, staring into Caruso's kindly face.

"Peter notified the FBI, and they have located Quincy in a motel on the outskirts of Bangor, Maine."

Henson stood, somehow restored to reality. "Thank you for your persistence, Caruso. That's damn good police work. We can take the helicopter and be there in forty-five minutes."

***

High above the clouds, the sky sparkling with a blanket of stars, Henson and Caruso sat in the rear seat of the police helicopter, a sergeant in the passenger seat beside the pilot. Caruso pulled his cellphone from his inside coat pocket, and searched his emails. Prissy had posted the fingerprints. He tapped Henson on the shoulder and extended the cellphone which shone brightly in the dimness of the cabin.

"Prissy analyzed the fingerprints in her files and got a perfect match on two key fingerprints. Quincy is our man all right."

Henson bobbed his head as his eyes pursued the prints, "The man must have needed money awfully bad."

"Yeah, and Dr. Kosaku bears some of the guilt, perhaps the greatest, because he devised the plan that paid Quincy."

The pilot answered a call from the local sheriff who reported that the FBI had taken Quincy into custody after he had discovered the stakeout and ran to his car. Henson instructed the pilot to inform the sheriff

290

that the prisoner was wanted in New York City, and NY State had jurisdiction. If Maine needed extradition rights it was up to the FBI.

They finally landed near the motel amid flashing lights from the police cars. Peter Meirs met Caruso and Henson and escorted them to the car where Quincy was handcuff. The arresting agents of the FBI had read him his Miranda rights, but he had no lawyer. Caruso flipped open his cellphone and called Jenny Lynn. He briefly discussed the situation, and she agreed with him that they would hire a lawyer to represent Quincy. Caruso closed the phone, and tugged the shoulder of the FBI agent in charge. He gave him his business card, and told him that Quincy would have a lawyer and had no need of a public defender.

Henson walked over to the car where Quincy sat handcuffed in the backseat. He released his gaze, and extended his handshake as he placed his other hand over Caruso's grasp. "Dr. Caruso, I owe you and your sidekick a great deal. If you are ever in the area again, we can take out the boat."

"I'll hold you to that, Randolph, but most of the glory goes to Prissy."

Use of his first name no longer bothered Henson. "Yeah, she's good at her work, and Brad needs her."

"Listen Henson, I like for you to come to my wedding—all right?"

Both bushy eyebrows elevated. "I wouldn't miss it, and congratulations, Jenny Lynn is a good woman, and she's been through a lot."

Caruso released the handshake. If only Henson really knew, would he be surprised. And very soon Caruso was headed for the surprise of his life.

Chapter 60

CHIP HURRIED DRESSING in his apartment's bathroom for a rescheduled dinner date with Marcy. Just as soon as he had returned from Peru, he had spent the day filing reports with the police. Dr. Caruso had been quite helpful in finalizing the details with Henson, and now he understood his mother's fondness for this interesting man; she needed a friend, and she had his approval of Caruso. But was Marcy the woman for him, or rather, was he the man for her. He remembered how careless he had been with the affections of Beverly, and he wanted a clean start, the same new day he'd given Mario.

He took a final look at his tie: was it the correct match with his black jacket, had he brushed his teeth. It struck him that he'd never had such thoughts when he crashed in Beverly's apartment, but he wasn't sure it meant anything. Yet he was soon to realize that it did.

He donned his dinner jacket, and took his final look in the mirror, then walked to the door. As he stepped out on the front porch, he noticed the air was cooler than when he drove home from the Midtown Precinct. He walked back into the house and took an overcoat from the hall closet.

His mind suddenly flashed back to Peru, and little Mario. Had he done the right thing by placing the money in a trust fund? Somehow this charming lad reminded him that he and his father were not close, too busy for much attention, and that's why he relished the summers when they flew off to Tanzania and Mount Kilimanjaro. His father was a good physician, and he respected him greatly, but his mother was the person of the family he adored. He sighed deeply and made a decision: if the bank claimed the money then he'd replace the money himself. In fact, he'd place a call to the hotel in Peru today and arrange for Mario's flight to New York. He'd speak with Dr. Caruso, perhaps he had connections to . . . and then it hit him squarely in his vanity. He'd adopt Mario, that is, if he wanted. Oh well, he sighed antagonized by his indecision, perhaps Marcy had some suggestions, he thought.

Suddenly he decided to call Peru at that very moment. He opened his cellphone and called the overseas link to the hotel desk in Peru. The registrar answered and told him that Mario was greatly excited about the trip. Chip informed the registrar that the American ambassador would pick him up at the hotel and fly with him to New York, and he would meet Mario at the airport on the coming weekend. Meanwhile he'd speak with Dr. Caruso; perhaps he had suggestions as to the legality of the immigration policy.

***

As he parked in front of Marcy's apartment, he noticed she had already opened the front door and stood in the threshold. Was he late, he thought, or was Marcy overly hungry? Chip hurriedly jumped from his seat and rushed to the door. She took his hand as he approached.

She smiled ruefully. "You're late, doctor—just suppose I needed an appendectomy, what about that, Chip?"

He shook his head embarrassingly, and realized how he had depended on Beverly even for his

293

appointments. "Gosh, I'm sorry, guess I simply lost track of time—can you ever forgive me, Marcy?"

Her smile said she had forgiven him already. Yet in her mind she realized how much men needed someone who kept them on track. And she was determined that she would not be another Beverly.

As he walked Marcy to the car, he sensed her perfume, so feminine, so alluring, and so breathtaking. She felt his eyes on her neck, and realized her plan was working. Somehow on this very night, she'd convince this man to kiss her, and then the door of his busy life would open to her intuition. She turned her head and gazed into his face.

"Where are you taking me to dinner?"

He seemed startled by her direct approach. "Ah, thought we'd dine at Tiffney's."

An eyebrow arched. "Thank you, Chip, that's so nice."

The drive seemed limitless; the lure of her beauty had stopped the clock in his mind. He drove as if drawn by a spider web luring him helplessly into its lair. Finally the restaurant appeared in the glowing lights of Fifth Avenue. He handed his key to the valet, and walked around the hood and opened the passenger door. Marcy took his hand, and he saw the warm glow of her pristine face, felt the gentle touch of her slender fingers that sent a tingle through his heart. She swung her long legs from the floorboard. A sudden urge gripped his emotions, and he pulled into his arms and planted a kiss on her ruby lips.

A tear rolled down her powered face. "That's so sweet, Chip!"

He stored his romantic thoughts, and closed the door.

As they walked toward the restaurant front door, Chip taxed his mind for words—truthful words of his hidden emotions. Yet, all he thought of was the response of Marcy when he would explain his intended adoption of Mario.

***

As they sat at a reserved table in the restaurant, a lit candle cast a mystic light over the chiseled checks of Chip's face. Marcy took a swallow of red wine, and slowly sat it on the maroon tablecloth, marveling at how well it matched his white tuxedo. She stared at the handsome man seated opposite, his tanned hand raking a fork across a napkin. Her heart smiled.

"Well, you haven't told me about the trip to Peru."

His head lifted, and he dropped the fork. It was the opportunity he'd hoped. "Morris McKinley died of snakebite."

Her alluring eyes swelled. "Oh, how grotesque."

"Yes, isn't it?" he replied, his mind searching for words. "Listen Marcy, there's something I've been meaning to ask you."

A sigh, almost a gasp, stuck in her throat. Was her plan that effective?

"Yes Chip?"

"McKinley favored a native boy in Peru–he's quite a charming little guy of twelve. I wondered what you think . . . if I should adopt him, that is, bring him to New York."

She leaned back in her chair, brought her napkin to her open mouth. "Why, Chip that's a marvelous thought."

"Well then, if that is settled, how would you like to meet Mario at the Kennedy Airport next weekend?"

Her eyes blossomed. "I'd like that very much. We both could meet him. I'll reschedule my duties. I'm so excited," she said, as she clasped his face between her slender hands, and kissed him.

His tight muscles released; the calmness of the starry night, the warm splendor of candlelight, the

295

openness created by red wine, the warmth of her kiss still tingling on his lips, all captivated his relaxed mind. No thoughts of the clinical trial, no distraction by nurses and phone calls.

Before he could reply, a waiter laid two platters before them, and refilled their wine glasses. He draped a towel over his arm, and bowed.

"Will you have something else, sir?"

Chip thanked him and replied, "Could we have the evening soup, please?

He bowed. "Yes, right away, sir."

As the waiter left, Marcy's right eyebrow suddenly arched. "Will the authorities allow you to keep Mario?"

Chip focused his mind on the sharp point of the question. "As to legality, I'm not at all sure now that you mention it, that is, a single man making a plea for adoption—doesn't sound too impressive, does it."

Marcy saw her chance and dove into it. "There is one way."

Chip released the wine glass from his lips. "And what do you suggest?"

The question must be answered even if she had to answer it herself. "A married couple could legally adopt," she tartly replied.

Chip gulped as he sat his wine glass down. Did she mean what she just said, he thought, or was she suggesting marriage! Dare he jump into this arrangement too soon? He swallowed, and took another gulp of wine. The brain power was immense; his thoughts tracing all that had gone wrong with Beverly. Should he inquire of his mother, or was that too childish. Again he swallowed. His mind had no answer, and he stood naked without an answer. He dared look into her face, the perfume overpowering the smell of the wine. A sigh brought both his hands to an outstretched position, and he took her hands.

"Marcy Curtis, will you marry me?"

There! He had said it, he told himself as he released one hand and reached into his coat pocket, and then placed a tiny felt covered box on the table.

Marcy sat dumfounded, a glaze over her eyes. All she had planned, all she had thought about, and Chip was already prepared to take her in marriage. He'd upstaged her; what a glorious thought. Tears cascaded down her cheeks as she opened the box.

Sparkling in her wet eyes was a two-carat diamond ring. She dropped her arms beside her chair; a feather would knock her over.

Chip took the ring from the box as she removed her glove, and nervously extended her left hand. He slipped the ring over the finger that Marcy had caressed for a month after meeting him, as she wiped the tears from her eyes.

"Yes, Chip. I will marry you, and we will adopt Mario."

Chapter 61

THE NEWS OF Chip's proposal to Dr. Marcy Curtis had reached the hospital grapevine, and finally Jenny Lynn knew it too. It was the moment that Dorothy Millhouse had predicted—the time Jenny Lynn must tell the truth. Jenny Lynn had scheduled dinner in the hospital café tonight at Dorothy's insistence; Robby and Chip had agreed to attend, both men destined for marriage. Marcy had finished her rounds, and came down the elevator, and went directly to Jenny Lynn's office. She found her future mother-in-law seated at her desk, doodling on a notepad.

"Hello, Jenny Lynn, thought I'd check in and see if you needed help with the dinner tonight."

She dropped her pen. "Oh, thank you Marcy, I wanted to talk with you before the dinner," and realized she had no the time for girl talk. "My dear, sweet Marcy, how thankful I am that you love Chip. I thought he'd never leave that laboratory, unless someone cared, and you have come to my aid, and his salvation."

Marcy threw her arms around Jenny Lynn. They hugged for what seemed an hour, yet it was only a blissful minute. Marcy whispered in her ear. "From the day I first saw Chip, I knew he was the one."

They parted at arm's length. "That's the way I think about Dr. Caruso," Jenny Lynn admitted, uninhibited by the moment of honesty.

Marcy's face wrinkled. "How's that," she wondered.

Release of the intimate information harbored in her heart for so many years, had relaxed her muscles as if a great weighty load suddenly lifted from her chest.

"Sit down Marcy."

As they sat, Jenny Lynn consulted her mind, yet all she visualized was Dorothy's confident words. Finally she gazed into Marcy's expectant eyes.

"Dr. John Cook is not Chip's biological father."

Marcy's eyes gapped wide open. The news was explosive, but not detrimental to her love for Chip. He was his own man, and she loved him for who he was, not for his ancestry.

"Would you care to tell who Chip's father is?" she ventured.

Jenny Lynn finally recovered from the price of admission. "You deserve to know, but I must hold you to secrecy, until I have the nerve to tell Chip."

Marcy gathered her into her arms, and kissed her on the cheek. "Your secret is safe with me, darling."

***

It was late afternoon and Chip had just time enough to visit with Debbie before dinner with his mother. Chip found Clyde in the room when entered. Debbie gave her best smile since she had been admitted into the program.

Chip nodded at Clyde as he approached the bed. "Well, Debbie, how did you like the steak I sent up yesterday?"

Her eyes blossomed like a tiny rose. "Swell, Dr. Cook, loved every bit of it."

"Well, I know Mr. O'Malley will like to hear that, he cooked it especially for you."

"Gee, thank you Dr. Cook."

Chip studied Debbie's chart for a long moment, and hooked the clipboard on the end of her bed.

"Clyde, I think we have some good news."

Clyde eyes blinked, the message unbelievable, as his hand rubbed his unshaven facer. He'd relieved his wife who usually stayed in the room. "I wish Martha were here," he whispered, a glaze burring his eyes.

Chip placed his hands on the shoulders of his best friend. "Well, why don't you call her? Debbie's cancer is in remission!"

Debbie leaped up in the bed and bounced on the mattress. Clyde sat down, tears flooding his tired eyes. He looked up at Chip as he wiped his eyes with the back of his hand.

"How can I ever repay you my friend?"

Chip gripped his bottom lip between thumb and index finger. "Well there is a way. You can stand in as best man at my wedding."

Clyde recovered his composure. "So you finally popped the question. I've talked with Marcy several times when she treated Debbie. I saw it in her eyes, heard it in her voice—the girl loves you Chip."

"How I know it brother, now call Martha, and give her the good news."

While Clyde phoned his wife, Chip took Debbie's hand. "Soon, young lady I will bring you a playmate. He's just about your age, his name is Mario."

A youthful smile spread across her face.

***

Chip left Debbie's room, finally finished his rounds, and then took the elevator to the clinical trial floor. He found Marcy at her desk pouring over the records of the clinical study. He tip-toed behind her chair, and placed his hands over her eyes.

She dropped her pen and placed her hands over his. "You can't fool me, doctor. I heard you coming."

He spun her chair around, gripped her hands, and lifted her to her feet. All in one motion while kissing her, and rubbing his hands up and down her back. "Got some good news today," he whispered in her ear.

"What's it going to cost me?" she replied nudging her nose on his cheek.

"Dr. Caruso has arranged for the ambassador to fly Mario to New York. He'll arrive next week."

A contagious smile spread her ruby lips. "How nice of Dr. Caruso, he certainly is a kind man."She pushed back from his arms, gazed into his sparkling eyes. "You've got something else on your mind, and see it in your hazel eyes."

He nodded. "Debbie cancer is in remission!"

In her heart she now realized that she would not attend the dinner. It was Jenny Lynn's moment.

Chapter 62

THE DINNER HOUR had finally arrived, and Jenny Lynn looked at the arrangements that the hospital waitress had provided for the dinner in a side room. Flowers at center table, an urn of coffee, brewed just as Robby preferred. See marched around the table somewhat nervous, recalling all that Dorothy had said, how simple it was: *just tell the man he has a son*, she quoted in her mind. Not that she disbelieved it, only that she had found it difficult to speak. But since she had told Marcy, perhaps that practice was the confidence she needed. And how she hoped it was true.

She waltzed over to the coffee urn and poured a cup of coffee hoping the caffeine calmed her nerves. The spout leveled her cup. Suddenly she heard noises in the corridor as the elevator closed. She turned and Dr. Robert Caruso strolled in with Chip following in his footsteps, broad smiles on each face. But where was Marcy, she thought?

She took a deep breath and sighed. The moment had at last arrived. Robby pushed her chair up to the table, and Chip took her hand and kissed it. This woman was his delight, the mother he cherished, and this man had become his valued friend, but the reason for this dinner together was unclear, and he was anxious to hear his mother's reaction to the questions that had stirred in his mind since his father was murdered and this Dr. Caruso had appeared out of nowhere.

The platters were served, the drinks provided, but no one seemed hungry. Jenny Lynn finally raised her glass of ice tea.

"A toast," she announced, her mind searching for words.

Caruso raised his glass. "Let me tell him, Jenny Lynn."

Chip's face flushed. Tell me what, he mused.

Caruso stretched his glass toward Jenny Lynn. "Chip, your mother, and I would like to announce our wedding."

Jenny Lynn smiled. It wasn't the toast she had planned, but Robby came to her rescue as usual.

Chip stood and raised his glass. "Mother, I agree with your choice."

"Then it's settled," she said. Now how would she tell him the truth?

Desert was served after a long discussion, while they had picked at the food with forks. The cake and ice cream had a different reaction on their appetites. They quietly munched on the desert, the silence disturbed only by the clank of forks against the plates. Jenny Lynn laid her napkin by her plate.

The moment had arrived.

"Chip, my son, I must tell you . . . " she swallowed.

Chip placed his fork on the plate, and raised a cup of coffee with his elbow on the table, the cup perched precariously his index finger. "Tell me what, mother."

She took a deep breath, the silence deafening.

"Dr. Caruso is your biological father!"

Chip's eyes bulged, coffee spilled down his front as he pushed back his chair.

Caruso gulped, sat his cup on the table grasping his chest as he slid to the floor.

Chip forgot his personal thoughts as his professional training seized his actions. He bounded to the floor and pounded on Dr. Caruso's chest as Jenny Lynn called for a crash cart.

Chip tore open Caruso's shirt, placed his ear on his chest. His heart was beating erratically. A crash cart rolled up beside him and he rubbed gel on the defibrillator pads and positioned them on Caruso's chest, calling, "Clear!"

The monitor was a straight line. Chip screamed, "240, Clear!"

The silence agonized alarmingly. "Dad wake up . . . wake up!"

Then the sound of blips on the monitor zigzagged across the screen: His heart was beating!

Jenny stood with her hands over her mouth, tears rolling down her cheeks, her heart thumping wildly. Attendants came in with a stretcher and gently placed Caruso on the canvas. Jenny Lynn followed them to the elevator, sniffing.

***

Chip placed his stethoscope on Dr. Caruso's chest, and moved it over a few areas of his heart. He reached up and repositioned the overhead light in the OR. Finally he released the stethoscope, and it dangled around his neck. He stood, not perplexed, only confused. It seemed that Caruso's heart was strong as a bull. He asked him to speak for a test of the effects on his speech functions.

Caruso spoke softly. "After the wedding, we'll make it official young man," he replied with only a hint of slur in specific words.

Chip faced Jenny Lynn with an arched eye. "Well I think my father is okay; a little therapy will do just fine," he said with a broad grin.

Jenny Lynn's heart skipped a beat as she took Caruso's hand. A tear formed at the edge of each of her blue eyes. "Darling, you gave us a scary moment."

He squeezed her hand. "And how do you think I felt?" he smiled. "How long had you intended to keep this news bottled up in your mind?"

Tears slowly streamed down her rosy cheeks. "It's a long story—we'll have lots of time to discuss it."

He managed a wink and turned his gaze toward his newfound son. "Well, Chip, I guess we have a lot of catching up to do."

Chip smiled broadly. For some reason he felt clean as if a load of clinical waste suddenly fell from his body—waste that had destroyed his social life. Perhaps medical science required no emotion, only dedication. And yet, he had learned that emotion was viable medicine when truthfully administered. He'd seen that in his newfound father. Yes, he thought, this was a man from whom he intended to learn.

***

Marcy received the news of Caruso's Transient Ischemic Attack (TIA) and was briefly horrified until Jenny Lynn had called and told her the complete story. And while they talked, Mary asked for her forgiveness by not coming to the dinner. Jenny Lynn only hugged her. They made plans for a double wedding.

Marcy voiced her concerns about the legality of adopting Mario. Jenny Lynn wondered if she and Chip may be forced to marry in Peru as touring Americans; she surmised that adoption was legal in that circumstance. She decided to discuss it with Robby, who was close to the President and the immigration procedures.

When Jenny Lynn returned to the hospital ward she found that Robby had wheeled his wheelchair into her office. As they drank coffee–the method which always put Robby in a good mood–she brought up the subject of Mario's adoption. Robby heard every word she'd spoken, but only smiled in reply. Finally the silence became too much for her.

"Well! Don't you have an opinion?"

Caruso understood Jenny Lynn better than his own habits. "I've already discussed this same matter with Chip earlier this morning."

"Well, I should have expected something like that from you," she smiled and took his hand, pressed it against her cheek. "Guess I'll have to remember your intransigent opinions since you've been in politics."

He grinned. "Let's just say it's the best way to get something done."

She cocked her head. "Just want have you done?"

"I've arranged for Mario's transport to New York through the Peruvian Embassy."

"That's legal?"

Caruso's hazel eyes said 'yes' as he opened his cellphone and punched a code. "Mario has one thing going for him, he's a minor." The cellphone buzzed with a connection. "Let me speak with Cavits, please." While he waited he smiled at Jenny Lynn, and blew her a kiss. "Cavits . . . Dr. Caruso, like to ask a question . . . Something like that . . . Peter Meirs has already discussed it, huh . . . I see . . . Thanks for the info."

He closed the phone. "Cavits thinks the green card is sufficient."

"So Peter Meirs is at it again–the old softy," Jenny Lynn grinned. She had been through several scraps herself long ago, and Peter had been a great friend.

"Yeah, can't do without him—it's like he's grafted to my hip."

"I miss him, too. You know I remember more than you realize."

His eyes glazed over with moisture. "Honey, there is so much for me to forget now that you are here beside me."

Robby's words touched her heart as memories coursed through her mind anew. She must talk with Dorothy soon, she thought.

The President called to offer his blessings. Jenny Lynn asked Robby to let her speak to the Chief Executive. He smiled, placed the phone to his ear.

"Mr. President, my wife would like a word . . . yes Sir."

Robby gave her the phone. "He says he'd like nothing better than to talk with you, darling."

Jenny Lynn cleared her voice. "Mr. President . . . yes Mr. President. I would like for Robby to retire and settle down with me. How do you feel that, Sir . . . I see, and then may I reserve the right to consider your proposition . . . Thank you, Mr. President . . . Good bye, Sir."

She handed the phone back to Robby. The blank on his face told her he wanted to know what the President said.

Jenny Lynn took his hand. "The President gave you three months off with paid leave. After that he agreed that he and I would consider my proposal."

"Is that all?" he smiled giddily.

She cocked her head. "Not exactly —he promised to attend the wedding."

The group erupted into robust laughter. Marcy and Jenny Lynn suggested that they all go over to O'Malley's and celebrate. As they drove over to Stanton Island, Jenny Lynn opened her cellphone and punched a code.

"Dorothy! I want you to come over to O'Malley's. We are having dinner. After all that you and I have gone through, I thought it only proper to have you by my side. Besides, I want to tell you in person what a good friend you have been."

The President did arrive for the wedding, after scheduling a meeting with the governor of New York. It was his chance to finally meet Jenny Lynn. Perhaps he knew more than Dr. Caruso about this woman. She would have tight reins on him. He knew this from his marriage. Women had a way of getting their way. Suddenly his aide whispered in his ear at the governor's meeting. He smiled.

"Dinner at O'Malley's, hum," he hummed. "Send an e-mail to Dr. Caruso telling him we would love to have dinner—wait, strike that! E-mail Jenny Lynn instead."